T0135062

Exact Design of Digital Microfluidic Biochips

Oliver Keszocze • Robert Wille • Rolf Drechsler

Exact Design of Digital Microfluidic Biochips

Springer

Oliver Keszocze
University of Bremen and DFKI GmbH
Bremen, Germany

Robert Wille
Johannes Kepler University Linz
Linz, Austria

Rolf Drechsler
University of Bremen and DFKI GmbH
Bremen, Germany

ISBN 978-3-030-08135-5   ISBN 978-3-319-90936-3   (eBook)
https://doi.org/10.1007/978-3-319-90936-3

Printed on acid-free paper

This Springer imprint is published by the registered company Springer International Publishing AG part
of Springer Nature.
The registered company address is: Gewerbestrasse 11, 6330 Cham, Switzerland

Preface

Many biological or medical experiments today are conducted manually by highly trained experts. Usually, a large laboratory requiring a lot of equipment is needed as well. This makes the whole process expensive and does not allow for very high throughput. This led to the development of automated laboratory equipment such as automated robots. These devices already allow for a high level of automation and integration. Unfortunately, laboratory robots are usually bulky (and expensive) and also use rather large amounts of liquids, which may be very expensive on their own.

To further reduce the size of laboratory devices, researchers investigated how to manipulate liquids at a nanoliter or even picoliter volume scale. This led to the development of microfluidic biochips, also known as lab-on-a-chip. The technical capabilities of microfluidic devices have been widely illustrated in the literature. An essential step for being able to actually make use of *Digital Microfluidic Biochips* (DMFBs) is to properly design (or synthesize) those. This process includes to take a medical or biological assay description, a biochip geometry, and further constraints and determine a precise execution scheme for running the assay on the biochip. As biochips grow in size and more complex assays are to be conducted, manual design of these devices is often not feasible anymore. Moreover, manual designs are often far from being optimal. Instead, high-quality design methodologies are required which relieve the design burden of manual optimizations of assays, time-consuming chip designs, as well as costly testing and maintenance procedures.

This book presents exact, that is minimal, solutions to individual steps in the design process as well as to a one-pass approach that combines all design steps in a single step. The presented methods are easily adaptable to future needs. In addition to the minimal methods, heuristic approaches are provided and the complexity classes of (some of) the design problems are determined.

By this, the book summarizes the results of several years of intensive research at the University of Bremen, Germany, the DFKI GmbH Bremen, Germany, and the Johannes Kepler University Linz, Austria. This included several collaborations—most importantly with the group of Prof. Krishnendu Chakrabarty from the Duke University, USA, and the group of Prof. Tsung-Yi Ho from the National Tsing Hua University, Taiwan. We would like to sincerely thank both colleagues for the very inspiring and fruitful joint work. Besides that, we are thankful to the coauthors of corresponding research papers which formed the basis of this book, including (in alphabetical order) Alexander Kroker, Andre Pols, Andreas Grimmer, Jannis Stoppe, Kevin Leonard Schneider, Maximilian Luenert, Mohamed Ibrahim, Tobias Boehnisch, and Zipeng Li. Furthermore, many thanks go to our research groups in Bremen and Linz for providing us with a comfortable and inspirational environment from which some authors benefit until today. Finally, we would like to thank Springer and, in particular, Charles "Chuck" Glaser for making this book possible.

Bremen, Germany   Oliver Keszocze
Linz, Austria   Robert Wille
Bremen, Germany   Rolf Drechsler
January 2018

Contents

Chapter 1
Introduction

Today, many biological or medical experiments are conducted manually by highly trained experts. Usually, a large laboratory requiring a lot of equipment is needed as well (see Fig. 1.1a which shows a typical laboratory setup). This makes the whole process expensive and does not allow for very high throughput. Furthermore, as human beings are no perfectly working robots, they are a source of errors, especially when many repetitive and monotonous steps are involved in a biological assay.

This led to the development of automated laboratory equipment such as the robots shown in Fig. 1.1b. These devices already allow for a high level of automation and integration, even though in many cases they only physically imitate the steps a human being would perform. Despite already significantly easing laboratory work, this still leaves room for improvement since those laboratory robots are usually bulky (and expensive).

To further reduce the size of laboratory devices, researchers investigated how to manipulate liquids at a nanoliter or even picoliter volume scale. This led to the development of *microfluidic biochips* (see Fig. 1.1c), also known as *lab-on-a-chip*. These are devices that automatically manipulate small amounts of liquids in order to perform (a subset of) the same experiments previously conducted in a laboratory. In addition to simply saving liquids, which may be expensive or difficult to obtain, smaller volumes can also result in shorter experiment execution times. In general, a higher throughput and sensitivity may be achieved.

The capabilities of microfluidic devices has been widely illustrated in the literature. Early works successfully demonstrate the applicability of biochips for multiplexed real-time *polymerase chain reaction* (PCR) [Liu+04] and colorimetric glucose assay for various bodily fluids [Sri+03]. In [Fai+07], different applications for biochips, such as massively parallel DNA analysis, real-time bio-molecular detection and recognition are presented. In [HZC10], protein crystallization for drug discovery and glucose measurement for blood serum are reported to have successfully been implemented. Another area where biochips are of interest is sample preparation [HLC12, Bha+17a, Bha+17b]. Using biochips, this tedious

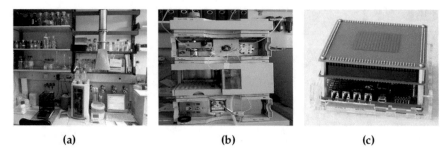

Fig. 1.1 Development in equipment size. (**a**) Laboratory. (**b**) Robot. (**c**) Biochip

process can be automatized to a high degree. In [Sis+08], biochips capable of executing different types of assays are used for point-of-care diagnostics. As has been pointed out in [Ali+17], biochips may be the future of easily accessible health-care. One scenario is to conduct on-site tests for diseases in remote regions. Besides that, also applications, e.g., for bubble logic [CYK07, PG07] or stochastic computing [HGW17] have been considered.

Motivated by this, different kinds of biochips and corresponding derivatives have been introduced which rely on different technologies. For example, *valve-based biochips* are composed of integrated *microvalves* [Hu+14, Mar+10], which are used to control the flow of liquids. Such biochips are made of materials such as glass, plastic, or polymers. Functionality, such as mixing liquids, is realized by fabricating corresponding channels at given positions. The microfluidic channels are used to transport the liquids to these positions. While, originally, such chips were rather static (similar to ASICs from conventional circuitry), in the meanwhile also more dynamic solutions have been proposed in terms of *Programmable Microfluidic Devices* (PMDs; similar to an FPGA from conventional circuitry [FM11, JBM10]). Figure 1.2 shows a valve-based biochip. While the chip itself is quite small in size, it still needs external hardware such as pressure sources. The overhead by the connectors to the chip itself is evident. The need for external hardware makes the operation of such a biochip "in the field" quite complicated.

In contrast, *digital microfluidic biochips* (DMFB) use an effect known as *electrowetting-on-dielectric* (EWOD) to actuate liquids [PSF02]. They comprise a two-dimensional electrical *grid* controlled by underlying electrodes and their electrical actuations. Using the actuations, an electric field is generated which allows to "hold" discretized portions of liquids, so-called *droplets*, on a particular cell within the grid. By assigning time-varying voltage values to turn electrodes on and off, droplets can be moved around the grid. Accordingly also certain operations can be realized, e.g. mixing by moving two droplets onto the same cell or heating by moving a droplet to a cell which comprises a heating device underneath. In order to actually induce a droplet movement, the droplet must have at least a minimal overlap with the electrode it is intended to be moved on. In order to make it easier to achieve this overlap, the edges of the electrodes are usually not manufactured as straight

Fig. 1.2 Physical realization of a flow-based biochip of the size of a dime [Whi06, Figure 1]

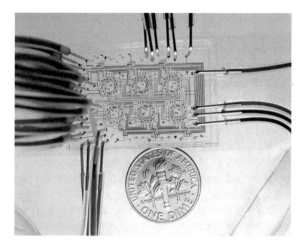

Fig. 1.3 Physical realization of a digital microfluidic biochip: zig-zag lined electrodes

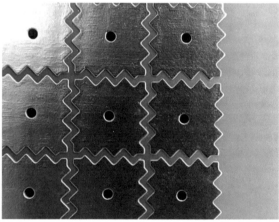

lines but as zig-zag lines, resulting in "interleaved" electrodes (see Fig. 1.3). Also this concept has been generalized so that eventually so-called *Micro-Electrode-Dot-Array* biochips (*MEDA biochips*, [Li+16, Che+11, WTF11]) resulted. Here, liquids are not controlled by single electrodes, but a *sea-of-micro-electrodes* is employed to allow for different droplet sizes and shapes.

Besides that, many further biochip technologies exist and/or recently received attention including, e.g., paper-based biochips as proposed in [WLH16a, WLH16b] or pressure-driven biochips (employing, e.g., the concept of *two-phase flow microfluidics*) as proposed in [De +12, Don+15, Don+14, De +13] which eventually resulted in a concept known as *Networked Labs-on-Chips* (NLoC, [De +12]).

In order to fabricate these biochips, there are several frameworks available. For example for DMFBs, open hardware solutions exist, e.g., in terms of the *DropBot* from the Wheeler Lab [FFW13] and the *OpenDrop* from GaudiLabs [OD]. The DMFB shown in Fig. 1.1c is a fully functional DMFB prototype that has been developed during the writing of this book. These devices can be manufactured in a very compact manner. The presented biochip is of dimension 11 cm × 11 cm × 6 cm and needs no further external equipment besides a common 12 V power supply. This is one major advantage of digital microfluidic biochips over flow-based biochips. No external pressure source or further equipment is required. Besides that, there also exist many successfully commercialized biochips such as Illumina's NeoPrep system [Illumina]. According to a report released by Research and Markets in June 2013, the global biochip market will grow from 1.4 billion in 2013 to 5.7 billion by 2018 [Market13].

However, in order to utilize these prospects, a corresponding biochip has to be designed (or synthesized) so that indeed the desired experiment is executed and, additionally, all constraints, e.g., with respect to the completion time are satisfied. This process includes to take a medical or biological assay description, a biochip, and further constraints, such as the maximally allowed completion time of the experiment, and use this input to determine a precise execution scheme for running the assay on the biochip. For the different kind of biochip technologies, a significant amount of corresponding automatic design methods have been proposed (see, e.g., [CZ05, Wan+17, Gri+17b] for valve-based biochips, [SHL16, Gri+18b] for PMD, [WLH16a, WLH16b] for paper-based biochips, or [Gri+18a, Gri+17c] for NLoCs). In this book, we will mainly focus on the automatic design of DMFBs (although many methods proposed here can also be applied for other biochip technologies).

Here, the objective of synthesis is to realize an experiment on the layout of the given biochip and within an upper bound on the completion time. To this end, the following design questions need to be addressed:

- Which modules shall be used in order to realize an operation? (*binding*)
- When (at what time steps) shall each operation be conducted? (*scheduling*)
- Where (on which cells) shall each operation be conducted? (*placement*)
- Which paths shall the corresponding droplets take in order to reach their destinations? (*routing*)
- Which electrodes can be grouped together in order to allow for a simpler control logic? (*pin assignment*)

These five steps are conducted in a two-stage design flow composed of an *architecture-level synthesis* (binding and scheduling) and a *physical-level synthesis* (placement, routing, and pin assignment) as illustrated in Fig. 1.4.

The input of a design problem usually consists of the following three parts.

Sequencing Graph   The sequencing graph describes the experiment in terms of operations and their interdependence. The sequencing graph implicitly defines the number of droplets and devices that are necessary. A sequencing graph is shown in the top left corner of Fig. 1.4.

Module Library   The module library describes what modules are available for realizing the necessary operations requested by the sequencing graph. A module library that can be used to realize the sequencing graph mentioned above is shown in the top center of Fig. 1.4.

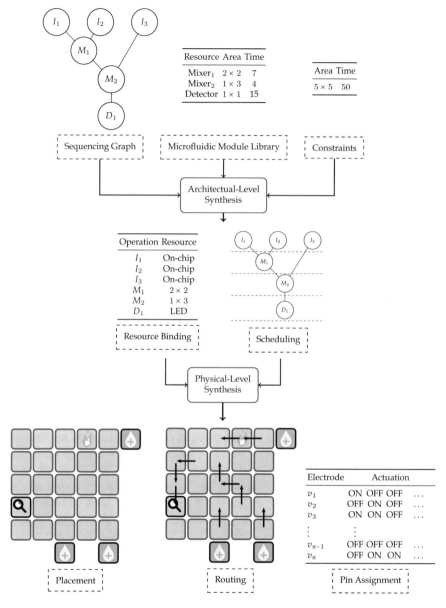

Fig. 1.4 Design flow for digital microfluidic biochips

Constraints   Usually, there are further constraints on the concrete realization of an experiment. These constraints can relate to the biochip itself (for example, an upper bound of available cells) or to the experiment (for example, an upper bound on the completion time).

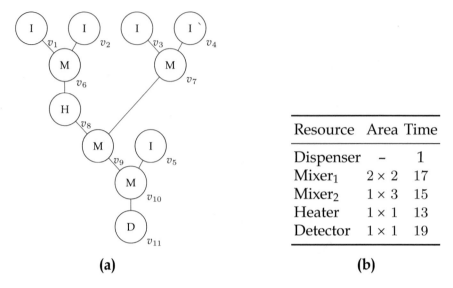

Fig. 1.5 (a) Sequencing graph and (b) module library for an experiment

A possible realization of the experiment as described in Fig. 1.5 is explained in the following example.

Example 1.1 Figure 1.6 illustrates the realization of the experiment on a 5×5 biochip. The visualization is shown in Fig. 1.6a while the precise timing information is shown in Fig. 1.6b. In the first time step, the droplets 1, 2, and 3 are dispensed. While the droplets 1 and 2 are mixed for 15 time steps in the lower mixer (indicated by the highlighted cells), droplet 3 is heated to its desired temperature for 13 time steps. The heated droplet 3 and the result of the mixing operation are then mixed for another 17 time steps. The resulting droplet is eventually analyzed by the detector in time steps 38–56. As can be seen, different fashions of modules are used for the mixing operation. The first mixer required 1×3 cells and 15 time steps, while the second one occupied a 2×2 cells over 17 time steps. Note that the time steps needed for the droplet movements are not explicitly listed in the table in Fig. 1.6b.

As biochips grow in size and more complex assays are to be conducted, *manual* design of these devices is not feasible anymore. Instead, high quality design methodologies are required which relieve the design burden of manual optimizations of assays, time-consuming chip designs, as well as costly testing and maintenance procedures.

For each of the five steps (binding, scheduling, placement, routing, and pin assignment), a number of dedicated, automated design methods have been developed, resulting in state-of-the-art solutions for *binding*, *scheduling* [Ric+06, SC08, GB12], *placement* [Che+13, YYC07, SC06a], *routing* [XC07, HH09, SHC06], and *pin assignment* [Xu+07, XC08, LC10], respectively (details of these steps as well

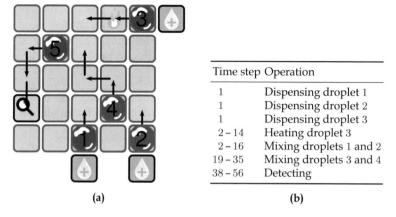

Time step	Operation
1	Dispensing droplet 1
1	Dispensing droplet 2
1	Dispensing droplet 3
2 – 14	Heating droplet 3
2 – 16	Mixing droplets 1 and 2
19 – 35	Mixing droplets 3 and 4
38 – 56	Detecting

(a)   (b)

Fig. 1.6 An experiment conducted on a DMFB. (**a**) Visualization of the experiment. (**b**) Timing of the experiment

as the corresponding methods are discussed later in this book in the respective chapters). While these approaches include very powerful and elaborated design solutions, they are mostly heuristic in nature. This means that the results do not necessarily come close to the optimal solution. A new heuristic method yielding an improvement of 10% can therefore either be insignificant or already close to the optimal results.

One of the contributions of this book is to analyze two design steps, namely routing and pin assignment, in detail. Theoretical results show that these two steps are **NP**-complete. These results have already been conjectured in the literature but never actually been proven. Having established the complexities of the problems, optimal, or exact, solutions using automated reasoning engines are presented. The use of such techniques is justified by the problems' complexity. The exact solutions to these problems already allow to determine solutions to interesting use cases. Furthermore, the exact results can be used to evaluate the quality of the previously proposed heuristic results. The scheduling and binding steps are not explicitly investigated in this book, as they can already be solved using techniques that are not specific to DMFBs.

As described above (see Fig. 1.4), the design problem consists of multiple steps. Usually, these steps are tackled separately and the individual solutions are combined to form the solution to the overall design problem. Even if the individual solutions are optimal with respect to reaction time, there is no guarantee that the overall solution is optimal as well.

To overcome this issue, a holistic *one-pass* approach that takes into account all steps of Fig. 1.4 at the same time is a necessary requirement for optimally realizing whole protocols. Consequently, another contribution of this book is an exact one-pass design approach. This approach guarantees the minimality of the overall solution to the design problem. The binding and scheduling steps, not considered so far, are implicitly handled by this one-pass approach. The proposed one-pass approaches still keep the pin assignment problem in a separate step.

Although ensuring optimality is usually computationally expensive, the exact synthesis approach is of great interest as, in addition to the evaluation of heuristics,

- it allows to determine smaller realizations than the previously best known and
- it allows to use the minimal realizations as building blocks for larger functionality.

While minimal solutions to the design problem are beneficial on their own, they also enable more sophisticated studies. One such study that is also conducted in this book is on the relation between the minimal biochip size and the minimal computation time.

As the pin assignment problem and the one-pass design problem are NP-complete, determining an exact solution may be too computationally expensive. Therefore, in addition to the exact solutions, heuristic approaches are presented. The remainder of this book is organized as follows.

First of all, to keep the book self-contained, Chap. 2 introduces the necessary technical background.

Chapter 3 deals with the routing of droplets. After the NP-completeness of the problem is proven, an approach for obtaining the optimal solution using an automated reasoning engine is presented. The routing solution is evaluated on a commonly used set of benchmarks.

In Chap. 4, the next step in the design flow, pin assignment, which is necessary to actually move the droplets after the optimal routes have been determined, is covered. Again, the NP-completeness is proven before an optimal solution is presented. Additionally, a heuristic framework for solving the pin assignment problem is introduced. The presented approaches are evaluated using results determined by the exact routing solution.

In Chap. 5, the results of the previous chapters are combined in order to solve the *pin-aware* routing problem. This problem is to minimize to necessary time steps as well as the number of pins to realize the routing solution. The pin-aware routing is shown to be very versatile. It will, for example, be used to optimize a given pin assignment. Furthermore, the solution is formulated in such a general fashion that is easily extended to, for example, route droplets on cells with a non-square shape or to consider cell degradation due to aging. It is shown that the ideas employed so far can easily be used in the context of a new technology for biochips: *micro-electrode-dot-array* (MEDA) biochips.

Chapter 6 finally introduces the one-pass design methodology. To the best of the author's knowledge, no such approach, optimal or heuristic, has previously been presented in the literature. In addition to the optimal approach, a heuristic solution is presented. The experimental results show that achieving high quality results using a one-pass heuristic in short computation time is possible. Still, the gap between the optimal and heuristic result is significant.

Finally, the book closes with a brief conclusion in Chap. 7. The findings of the book are summarized and open research questions are discussed.

Additional material used to create pictures for this book is presented in the Appendix. In Appendix A, a dedicated grammar for describing biochip configurations is introduced and in Appendix B a corresponding visualization tool is presented.

Chapter 2
Background

2.1 Microfluidic Biochips

2.1.1 Microfluidic Operations

Biological assays usually consist of a multitude of *operations* that need to be performed for the experiment to succeed.

To actually perform these operations, *modules* are used. These modules fall into one of the following two categories. *Physical modules* are realized by hardware that is built on the biochip. These modules are not reconfigurable in the sense that their positions are fixed. *Virtual modules*, in contrast, are entirely realized via electrowetting. This means that their position can be freely re-arranged, if necessary. For instance, a position that has previously been used to store droplets may become part of a mixing operation in the next time steps.

The module library specifies which modules can be used to perform which operation. A single operation may be realizable by more than one module. One example for this is the mixing process, where mixers of different sizes can perform the mixing in different numbers of time steps.

The following, non-exhaustive, list of physical modules gives an overview over what a DMFB is capable of.

Dispensers   Liquids to be used in the experiment are kept in *reservoirs*. Whenever required, a sample is taken from this reservoir and brought on the chip. For this purpose, *dispensers* for each liquid are physically added next to the outer cells of the grid. For each type of liquid considered in the experiment (for example, blood, urine, reagents), a separate reservoir and, hence, a separate dispenser has to be provided.

Sinks   If droplets are not needed anymore during the execution of an experiment, they might be removed from the grid (for example, in order to make room for other droplets and/or operations). For this purpose, similar to dispensers, sinks

© Springer International Publishing AG, part of Springer Nature 2019

O. Keszocze et al., *Exact Design of Digital Microfluidic Biochips*,
https://doi.org/10.1007/978-3-319-90936-3_2

are added to the outer cells of the chip. Since sinks are used for waste disposal only, no differentiation between types is necessary.

Heaters   Heating samples may be an integral part of an experiment. For this purpose, heating devices are placed below selected cells. Then, droplets occupying
this cell can be heated if desired. Heating may, for example, be necessary to
create perfect conditions for enzymes.

Detectors   At the end of an experiment, the properties of the resulting droplet shall
usually be examined. For this purpose, sensor devices are placed below selected
cells. Then, droplets occupying this cell can be analyzed with respect to different
characteristics such as color, volume, etc.

While physical modules always require corresponding devices built-in onto the
chip, some of the operations can implicitly be realized by the movements of droplets
instead of dedicated hardware. These operations include the following.

Mixers   Mixing liquids is an integral part of almost every experiment. Using
electrowetting, this can be realized by simply routing the droplets to be mixed
to the same cell. In order to accelerate diffusion, the newly formed droplet is
moved back and forth between several cells.

Splitters   Droplets resulting from mixing operations have twice the size of the
input droplets. To reduce them to normal size, they are split up. This can be
realized by simultaneously activating cells of the grid that are on the opposite
sides of the droplet. Then, the resulting forces split the droplet into two parts.

Storage   When a droplet is already present on the DMFB but not immediately
used, it needs to be stored somewhere. Storing a droplet is routing the droplet to
a position where it does not interfere with the rest of the protocol currently being
conducted. As this process does use cells and some time steps, some authors
explicitly model storing of droplets in their approaches.

Overall, modules allow for the realization of various operations to be performed
in laboratory experiments. Some of them are available in different fashions with
respect to the number of occupied cells and the number of time steps required for
their execution.

Note that, in practice, many further physical and virtual modules may be
available in a module library. But for the sake of simplicity, only the modules
reviewed above are used. However, the concepts and solutions proposed in this book
can easily be extended for further modules.

2.1.2 Fluidic Constraints

The issue of unintended mixing can occur as droplet routing significantly differs
from classical wire routing with its static, non-crossing routes. In the classical wire
routing one simply has to keep a certain distance between the wires and then can
be sure that no mutual influence will occur. It is possible to try to find such kind

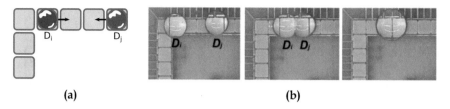

(a)   (b)

Fig. 2.1 Unintended mixing of droplets, static case. The pictures in (**b**) are taken from [SHC06, Figure 3]. (**a**) Two droplets moving to adjacent cells. (**b**) The two droplets come into contact and merge into a single big droplet

of routes on biochips as well, but that would completely ignore a core feature of biochips: While the droplets' routes themselves are static, the droplets' positions are not. This dynamic aspect allows for routes to cross each other or even partially overlap as long as droplets are never located too close to each other at every single point in time. Taking this into account, it is possible to determine shorter routes. One obvious situation that has to be avoided in order to prevent unintended mixing is having multiple droplets on top of the same cell. But as has been mentioned in [Böh04] and thoroughly analyzed in [SHC06], this is not sufficient. In addition to the case already introduced in [Böh04], the authors of [SHC06] identified another situation that needs to be taken care of. In the following, the notion of *fluidic constraints* taken from [SHC06] is adopted.

Consider the situation depicted in Fig. 2.1a. Two droplets are to be routed to directly adjacent cells. As droplets need to have some overlap to the neighboring cells in order to move there, the two droplets will come into contact with each other and merge (see Fig. 2.1b). This issue of unintentional mixing is captured in the *static fluidic constraint* stating that no two droplets must be adjacent to each other in any given time step. The cells around a droplet that must not be entered also include diagonally adjacent cells and will be called *interference region*.

Figure 2.2 visualizes the electrodes reachable by a droplet as well as the electrodes in the interference region.

Interestingly, the issue with neighboring droplets merging is not restricted to droplets that are adjacent in the same time step. As has been shown in [SHC06], entering a cell that has been in the interference region of another droplet in the *previous* time step can also be sufficient for droplets to merge (see Fig. 2.3). This issue is captured in the *dynamic fluidic constraint*. As can be seen, the interference region does not consist of the horizontally and vertically neighboring cells only. The diagonally adjacent cells also have to be taken into account.

The fluidic constraints do not apply to droplets that are intended to be mixed at some point in the conducted experiment anyway.

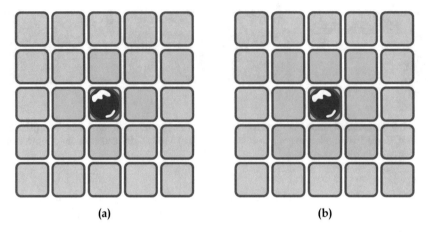

(a)   (b)

Fig. 2.2 Reachable electrodes and interference region for a droplet. (**a**) The electrodes reachable by the droplet in the center of the biochip. (**b**) The interference region for the droplet in the center of the biochip

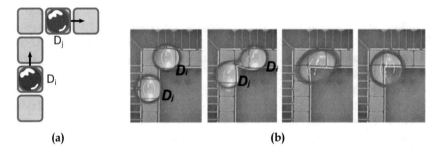

(a)   (b)

Fig. 2.3 Unintended mixing of droplets, dynamic case. The pictures in (**b**) are taken from [SHC06, Figure 4]. (**a**) One droplet moving into the previous interference region of another droplet. (**b**) The two droplets also come into contact and merge into a single big droplet

2.2 Discrete DMFB Model

In general, all physical aspects such as the voltage needed do drive droplets and the precise movement speed is necessary for understanding the DMFB technology as such. When trying to perform computer aided design for these chips, the level of detail is actually a hindrance. From the design perspective, designers are only interested in the fact that a biochip conducts a given assay and assume that the device is working correctly. For this, a high-level view on biochips is used that abstracts away implementation details that are not necessary for the design process. This is similar to the design for conventional circuitry where, e.g., voltages/currents and transistors are, respectively, abstracted in terms of 0/1 and gates, respectively. Also in the design and simulation of microfluidic devices, corresponding abstractions are rather common (see, e.g., [Gri+17a]). In the following, a model for DMFBs as used in this book, is introduced.

2.2.1 Geometry of the Biochip

The physical layout of the cells, called *grid*, is described by means of an undirected graph $G = (V, E)$. The vertices V correspond to the cells and the edges E model the possible droplet movements between the adjacent cells. As the precise graph formulation is very technical, it is used for proofs only. The rest of this book uses a more convenient notation. The vertices are called *positions*. The grid of a biochip is then described by a set of positions p denoted by $\mathcal{P}$. The positions are identified by Cartesian coordinates with the origin $(0, 0)$ in the lower left corner, that is, $\mathcal{P} \subset \mathbb{N} \times \mathbb{N}$. The edges are not explicitly stated but the set of reachable positions for a given position p is denoted by $N(p)$. It will also be called the *neighborhood* of p. In order to model that a droplet can wait on a position, this set always contains p itself.

Example 2.1 Figure 2.4a displays a biochip with a 4×4 grid (some cells are labeled for the sake of easy reference). Droplet movement is possible in horizontal and vertical directions. The set of reachable positions for $p = (0, 0)$ is given by $N(p) = \{(0, 0), (0, 1), (1, 0)\}$. All other positions are not reachable.

There is no technical restriction for biochips to be of such a regular "rectangular" shape. Biochips can also be manufactured in a way that the cells do not cover the entire PCB as shown in Fig. 2.4b.

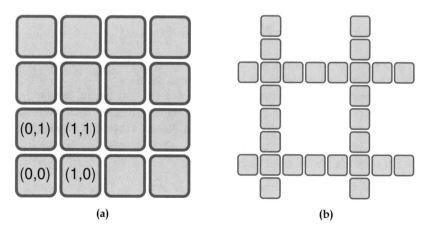

(a)   (b)

Fig. 2.4 Examples for different biochip grid geometries. (a) Rectangular 4×4 biochip consisting of 16 cells. (b) Non-rectangular biochip consisting of 28 cells

The positions that belong to the interference region of position p are denoted $N_I(p)$.

Some positions on the biochip may be blocked. This can be due to some operations being currently performed on these positions. A blockage means that no droplet is allowed to enter the blocked position. Such a blockage is denoted by b; the set of all blockages by $\mathcal{B} \subset \mathcal{P}$.

2.2.2 Droplet Movement

The most important abstraction is the assumption that a droplet moves from cell to cell in unit time. This unit time is identical for all droplets and the movement is coordinated to happen synchronously for all droplets used in the experiment. This allows to divide the execution time of an assay into discrete *time steps*. The time step is denoted by t and ranges from 1 to some upper bound T.

In this work, a droplet is denoted by d. The set of all droplets used in the currently investigated design is denoted by $\mathcal{D}$. These droplets are described using a unique identifier.

In order to be able to describe intended droplet movement, the concept of *nets*, which is borrowed from the wire routing problem for conventional circuits (see, for example, [Alb01, PC06]), is used. Nets are one of the most essential notions when describing biochip functionality as all operations are based on droplet movements.

Definition 2.1 (k-Net) A *net* is a means to describe that a set of droplets should be routed from given source positions to a common target position. Formally, a net n is a tuple of the form

$$n = \left(\{(d_1, p_1^*), (d_2, p_2^*), \ldots, (d_k, p_k^*)\}, p^\dagger \right),$$

where $d_i \in \mathcal{D}$ defines which droplets are to be routed, $p_i^* \in \mathcal{P}$ are the corresponding source positions, and $p^\dagger \in \mathcal{P}$ is the common target position.

The droplets belonging to the net n are denoted $\mathcal{D}_n$. The general form is $\mathcal{D}_n = \{d_1, \ldots, d_k\}$. The net to which the droplet d belongs to is denoted n_d. To allow for a more concise notation, the source and target position of a droplet d are denoted p_d^* and $p_d^\dagger$ without an explicit reference to the net n which the droplet d is part of.

A net consisting of k droplets is called k-net.

Example 2.2 Consider the situation depicted in Fig. 2.5a. Droplets 1 and 2 are to be routed to a common target position $(1, 3)$ from positions $(0, 4)$ and $(2, 0)$, respectively. Droplet 3 is to be routed from position $(4, 1)$ to position $(3, 4)$. The corresponding nets are

$$(\{(1, (0, 4)), (2, (2, 0))\}, (1, 3)) \quad \text{and} \quad (\{(3, (4, 1))\}, (3, 4)).$$

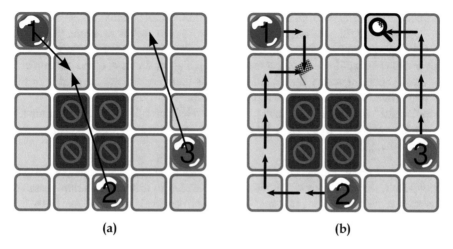

Fig. 2.5 Illustration of nets and routes. (**a**) Intended droplet movement for three droplets. (**b**) Droplet routes

It is important to note that the fact that two droplets have the same target position does *not* imply that they are in the same net. In fact, there can be multiple nets with the same target position.

The movements of a droplet $d \in \mathcal{D}$ are captured by the notion of a route.

Definition 2.2 (Route) A route r_d of length T for the droplet d is a series of positions $p_d^t \in \mathcal{P}$ for $1 \leq t \leq T$ such that $p_d^{t+1} \in N(p_d^t)$ for all $1 \leq t < T$. The droplet's position at time step t is written as $r_d(t)$.

Example 2.3 Consider the droplet movements depicted in Fig. 2.5b. The corresponding droplet routes are given by

$$r_1 = \big((0, 4), (1, 4), (1, 3)\big),$$
$$r_2 = \big((2, 0), (1, 0), (0, 0), (0, 1), (0, 2), (0, 3), (1, 3)\big), \quad \text{and}$$
$$r_3 = \big((4, 1), (4, 2), (4, 3), (4, 4), (3, 4)\big).$$

2.2.3 Electrode Actuation

As described in Chap. 1, droplets are moved by actuating the electrodes underneath the cells. Therefore, to actually move droplets, the actuations of the electrodes on the biochip must be known.

Different types of means for actuating cells on a biochips have been proposed resulting in basically three varieties of biochips:

Directly Addressing Biochips   These chips allow to individually actuate every single cell on the chip through a dedicated control pin (see, for example, [Xu+07]).
Cross-Referenced Biochips   In the scheme introduced in [FHK03], rows and
   columns are addressed only, activating the pin at the crossing of the column and
   row. These chips only have as many pins as there are rows and columns on the
   chip (that is, $W + H$ pins for a $W \times H$ grid).
Pin-Constrained Biochips   Another method to actuate cells is to employ a broad-
   casting scheme. This means that a single pin controls multiple cells (see, for
   example, [SPF04]).

While directly addressing all electrodes allows for a very flexible droplet
movement, the pure number of control pins needed to drive the biochip becomes
infeasible. As pointed out in [Xu+07], a biochip consisting of a 100×100 array
already needs 10^4 pins. This leads to a very complex wire routing, which can be
unacceptable for devices that are intended for a limited number of uses. In the same
work, the authors state that the physical realization of cross-referenced biochips
is expensive. Both the bottom plate and the top plate have to contain electrodes.
Furthermore, they are also unsuitable for high-throughput assays as the droplet's
movement is too slow. Due to the restrictions imposed on the droplet movements
by the addressing scheme, the routability of nets might not be as flexible as with
directly addressing biochips. While cross-referenced biochips have inspired a lot
of work and research (see, for example, [Yuh+08]), the negative aspects dominate.
Cross-referenced biochips will, therefore, not be considered in this work.

   Directly addressing biochips can be seen as a pin-constrained biochip using one
pin per cell. In the rest of this book, both types of biochips are treated the same, up
to the actual pin assignment. Depending on the choice of which pin controls which
cells, the reduction in overhead may be significant.

   To describe the actuation behavior of the cells of the biochip, the concept of
actuation vectors is used.

Definition 2.3 (Actuation Vector) During the execution of an assay, a cell can be
in one of the following states: *actuated*, *not actuated*, and *don't care*. Those states
are denoted by $1, 0$, and X, respectively. The set of these actuations is denoted by $\mathbb{A}$.
An element $v \in \mathbb{A}^T$ then describes the actuation behavior of a cell over T time steps.
Such an element is called *actuation vector*. The actuation vector corresponding to
position $p \in \mathcal{P}$ is denoted $v_p \in \mathbb{A}^T$.

Example 2.4 Consider the movement of a droplet as depicted in Fig. 2.6a. In time
step 1, the droplet is on position $(0, 1)$, which, therefore, needs to be actuated. The
horizontally and vertically adjacent cells must not be actuated as that would lead to
an unwanted droplet movement. In time step 2, the droplet moves to position $(0, 2)$.
To realize this movement, position $(0, 2)$ needs to be actuated while, at the same
time, the initial position $(0, 1)$ must not be actuated anymore. Note that positions
$(0, 0)$ and $(1, 0)$ still must not be actuated as this could, in the worst case, lead
to undefined droplet movement. In time step 3, the droplet is finally moved to its
destination at position $(1, 2)$. In this step, the actuation of position $(0, 0)$ and $(1, 0)$

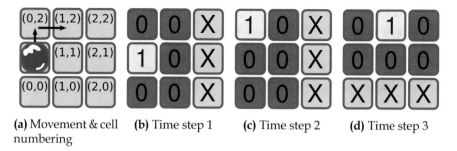

(a) Movement & cell numbering   **(b)** Time step 1   **(c)** Time step 2   **(d)** Time step 3

Fig. 2.6 **(a)** Movement of a droplet from position d to position b. **(b)**–**(d)** Actuation states of the cells at the corresponding time steps

is not important any more. The actuation of position $(2, 0)$ is not relevant for the droplet movement at all.

This droplet movement leads to the actuation vectors

$$v_{(0,2)} = (0, 1, 0), \qquad v_{(1,2)} = (0, 0, 1), \qquad v_{(2,2)} = (X, X, 0),$$

$$v_{(0,1)} = (1, 0, 0), \qquad v_{(1,1)} = (0, 0, 0), \qquad v_{(2,1)} = (X, X, 0),$$

$$v_{(0,0)} = (0, 0, X), \qquad v_{(1,0)} = (0, 0, X), \text{ and} \qquad v_{(2,0)} = (X, X, X),$$

which are illustrated in Fig. 2.6b–d.

Note that in some papers (for example, [HSC06]) it is argued that diagonally adjacent cells, when actuated, have no influence on the droplets. This less conservative approach would lead to more *don't care* values in the actuation vectors.

2.3   Reasoning Engines

One of the core techniques used in this book is the use of reasoning engines. The basic idea is to create a mathematical model of the problem that is to be solved and then let a powerful solving engine determine a valid solution.

This process usually means that variables describing the various entities of the problem need to be defined. After a domain for choosing the variables from (Booleans, Integers, etc.) has been determined, the variables need to be further constrained to faithfully model the problem. After this modeling is done, the model is given to a dedicated software. These tools (solving engines, or simply solvers) are capable of determining a valid assignment to the model or prove that no such assignment exists. This assignment then is the solution to the initial problem.

These formal methods are used to determine solutions that are guaranteed to have the smallest value for some optimization criterion. In this book the terms "exact" and "optimal" will be used to denote methods that give such a guarantee.

In the rest of this section, three commonly used solving techniques, two of which are used throughout this book, are explained.

2.3.1 Boolean Satisfiability

Given a formula φ in propositional logic, the satisfiability problem (SAT) is to determine whether there exists an assignment to the variables used in φ such that φ evaluates to true. For a thorough treatise on SAT, see [Bie+09]. Propositional logic already is powerful enough to model many problems, as can be seen in the following example.

Example 2.5 Consider the Boolean variables x_1, x_2, and x_3 representing whether a process i ($i = 1, 2, 3$) uses a restricted resource that only allows for one process to use it. To model that at most one process may use this resource, for each pair of variables that are chosen from the set of all variables, at least one must be false. The total SAT instance is given by

$$\varphi = (\neg x_1 \vee \neg x_2) \wedge (\neg x_1 \vee \neg x_3) \wedge (\neg x_2 \vee \neg x_3).$$

All possible solutions to this instance are

$$x_1 = \text{false} \wedge x_2 = \text{false} \wedge x_3 = \text{false}, \quad x_1 = \text{true} \wedge x_2 = \text{false} \wedge x_3 = \text{false},$$

$$x_1 = \text{false} \wedge x_2 = \text{true} \wedge x_3 = \text{false} \quad \text{and} \quad x_1 = \text{false} \wedge x_2 = \text{false} \wedge x_3 = \text{true}.$$

SAT solvers expect the input to be in *conjunctive normal form* (CNF). While all Boolean formulas have at least one CNF representation, it usually is more convenient to write formulæ in a more abstract form. In the rest of this book, for example, the term $x \implies y$ will be used instead of writing $\neg x \vee y$. This considerably improves the readability. The situation in Example 2.5 can be more naturally expressed as a sum of the form $\sum_{i=1}^{3} x_i \leq 1$. There are many papers on how to translate these *cardinality constraints* into a CNF, see, for example, [BB03, Sin05, ES06]. This book uses the approach from [Sin05]. Moreover, Z3 [DB08] is used to solve SAT instances.

2.3.2 Satisfiability Modulo Theories

Satisfiability Modulo Theories (SMT) allows to formulate decision problems in first-order-logic that are enriched with certain theories. In this book, the theory of integers is used. This allows to formulate problems in which alternative options need to be encoded as numbers.

Example 2.6 Consider the following graph.

The problem of *edge coloring* is to assign unique colors to all edges such that no two adjacent edges have the same color. This can easily be modeled using SMT when creating an Integer variable c_i for each edge e_i. The value of the variable c_i encodes the color for edge e_i.

The edge coloring problem that asks whether the edges can be colored using at most k colors can easily be formulated in SMT. The first step is to constrain the number of colors that are used when trying to color the graph.

$$c_1 \geq 0 \wedge c_2 \geq 0 \wedge c_3 \geq 0 \wedge c_4 \geq 0 \wedge c_5 \geq 0$$
$$c_1 < k \wedge c_2 < k \wedge c_3 < k \wedge c_4 < k \wedge c_5 < k$$

The mutual exclusions of colors can easily be determined from the graph and can be translated directly into the SMT instance as follows:

$$c_1 \neq c_2 \wedge c_1 \neq c_3 \wedge c_1 \neq c_4 \wedge c_1 \neq c_5$$
$$c_2 \neq c_3 \wedge c_2 \neq c_4 \wedge c_3 \neq c_5 \wedge c_4 \neq c_5$$

For $k < 3$, obviously no solution can be found. For $k = 3$, a solution can be determined. One possible variable assignment produced by an SMT solver is

$$c_1 = 2 \wedge c_2 = 1 \wedge c_3 = 3 \wedge c_4 = 3 \wedge c_5 = 1$$

The SMT solver used in this book is Z3 [DB08].

2.3.3 Integer Linear Programming

Even though the book itself does not employ *integer linear programming* (ILP) to solve design tasks, some of the cited papers do. To allow the reader to easily understand the approaches of these papers, it is introduced in the following.

ILP is similar to SAT and SMT in the sense that ILP also works on a model for which a valid instance is to be determined. In contrast to SAT and SMT, ILP is not a decision problem but a numerical optimization problem. Given n integer variables $x_i \in \mathbb{Z}$ with corresponding weights $c_i \in \mathbb{Z}$, the goal is to minimize

$$\sum_{i=1}^{n} c_i \cdot x_i$$

subject to the conditions $x_i \geq 0$ and

$$\sum_{j=1}^{n} a_{i,j} x_j \geq b_j$$

for all $1 \leq i \leq n$ and weights $a_{i,j}, b_i \in \mathbb{Z}$. For a thorough introduction see, for example, [WN99].

Example 2.7 Consider the distance matrix

$$\begin{array}{c} & \begin{array}{cccc} City\ 1 & City\ 2 & City\ 3 & City\ 4 \end{array} \\ \begin{array}{c} City\ 1 \\ City\ 2 \\ City\ 3 \\ City\ 4 \end{array} & \begin{bmatrix} 0 & 10 & 20 & 30 \\ 10 & 0 & 25 & 35 \\ 20 & 25 & 0 & 15 \\ 30 & 35 & 15 & 0 \end{bmatrix} \end{array}$$

where the entry $a_{i,j}$ is the time in minutes it takes to drive from city i to city j. The goal is to build fire stations in these towns in such a way that each city can be reached by the fire brigade in at most 20 min. The total number of fire stations should be as small as possible.

The corresponding ILP problem is given as follows. There are four variables x_i, representing whether a fire station is built in city i ($x_i = 1$) or not ($x_i = 0$). The term that is being minimized is given by

$$\min x_1 + x_2 + x_3 + x_4.$$

The constraints

$$\begin{aligned} x_1 + x_2 + x_3 &\geq 1 && \text{from city 1} \\ x_1 + x_2 &\geq 1 && \text{from city 2} \\ x_1 + x_3 + x_4 &\geq 1 && \text{from city 3} \\ x_3 + x_4 &\geq 1 && \text{from city 4} \end{aligned}$$

model the fact that there is a limited amount of time to reach each city. Every line makes sure that in all cities reachable from the given city within 20 min, at least one fire station is build.

At least two fire stations need to be built and one feasible solution is to build fire stations in cities 1 and 4.

Chapter 3
Routing

3.1 Problem Formulation

The routing problem for DMFBs is defined as follows:

Definition 3.1 (DMFB Routing Problem) The input of the DMFB routing problem consists of

- the biochip architecture, given by the set of positions $\mathcal{P}$,
- the, possibly empty, set $\mathcal{B} \subset \mathcal{P}$ of blockages, and
- the set $\mathcal{N}$ of nets.

The DMFB Routing Problem is to determine routes for all nets that do not violate the fluidic constraints and respect all the blockages present on the biochip. As a secondary problem, the minimization of route lengths can be considered.

One routing problem with a corresponding solution is illustrated in the following example.

Example 3.1 Consider the situation depicted in Fig. 3.1a. On a biochip of size 5×5, three droplets are to be routed. Droplet 3 should be moved from starting position $(4, 0)$ to an detecting device at position $(3, 4)$ while the other two droplets are to be routed to their common target at position $(1, 3)$. One possible solution of this routing problem using six time steps is shown in Fig. 3.1b.

3.2 Complexity of Routing

This section will analyze the computational complexity of the DMFB Routing Problem. The NP-completeness of the problem has already been conjectured in the literature. In [Böh04] the similarity to the NP-hard problem of moving multiple

© Springer International Publishing AG, part of Springer Nature 2019
O. Keszocze et al., *Exact Design of Digital Microfluidic Biochips*,
https://doi.org/10.1007/978-3-319-90936-3_3

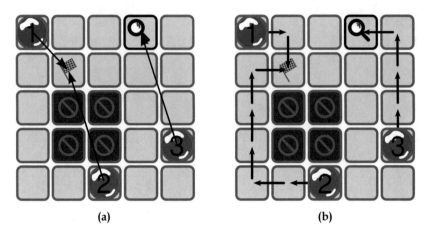

Fig. 3.1 Example for a routing problem with one possible solution. (**a**) Routing problem for three droplets: the biochip has a blockage of size 2×2 and a detecting device at position $(3, 4)$. (**b**) Exemplary routing solution using six time steps

robots has been noted ([CP08] uses a similar reasoning) while in [SHC06] the similarity to the Steiner Minimum Tree problem is pointed out. In this chapter, the $(n^2 - 1)$-Puzzle will be used in an explicit proof of the NP-completeness of the DMFB Routing Problem.

Before being able to formally analyze the complexity of the routing problem, as defined in Definition 3.1, it has to be formulated as a decision problem. The biochip will be modeled by a graph structure. The droplet routes then correspond to paths in that graph.

Definition 3.2 (DMFB Routing Problem as a Decision Problem) To formulate the routing problem as a decision problem, the formal graph model introduced in Sect. 2.2 is used. The interference region is modeled by another set I of edges.

Let $p_d^*, p_d^\dagger \in V$ be the source and target position of the droplet $d \in \mathcal{D}$. The net containing droplet d is denoted n_d.

The decision problem is then defined as follows. Given a maximal number of time steps $T \geq 1$, do there exist paths r_d in G for all $d, d' \in \mathcal{D}$ such that the assertions

$$(r_d(1) = p_d^*) \wedge (r_d(T) = p_d^\dagger) \tag{3.1}$$

and

$$\{r_d(t), r_{d'}(t - i)\} \notin I \tag{3.2}$$

for $i = 0, 1$ for droplets d, d' with $n_d \neq n_{d'}$ and $1 < t \leq T$ hold?

In the definition above, (3.1) ensures that the droplets start and arrive at the correct positions. The correctness of these paths is inherently ensured by the graph structure. That droplets which are not allowed to interfere with each other (that is, they belong to different nets) do not violate the fluidic constraints, is ensured via (3.2). The set I models the interference region. In the case of $r_d(t) = r_{d'}(t-0)$, the constraint enforces that no two droplets are on the same cell at the same time step.

The decision problem definition of the DMFB Routing Problem is illustrated in the following example.

Example 3.2 Consider again the example shown in Fig. 3.1. The problem description and the solution using the formulation of Definition 3.2 is as follows.

The graph $G = (V, E)$ is defined by the sets

$$V = \{(x, y) \mid x, y \in \{0, 1, \dots, 4\}\} \setminus \{(1, 1), (1, 2), (2, 1), (2, 2)\}$$

and

$$E = \{\{(x_1, y_1), (x_2, y_2)\} \mid (x_1, y_1), (x_2, y_2) \in V \land |x_1 - x_2| + |y_1 - y_2| \leq 1\}.$$

The interference regions around the positions is modeled by

$$I = \{\{(x_1, y_1), (x_2, y_2)\} \mid (x_1, y_1), (x_2, y_2) \in V \land |x_1 - x_2| \leq 1 \land |y_1 - y_2| \leq 1\}.$$

The set I is the set E with all the diagonally adjacent positions to be taken into consideration.

These sets, together with

$$p_1^* = (0, 4), \qquad p_2^* = (2, 0), \qquad p_3^* = (4, 1),$$
$$p_1^\dagger = (1, 3), \qquad p_2^\dagger = (1, 3), \qquad \text{and } p_3^\dagger = (3, 4),$$

describe the situation depicted in Fig. 3.1a. The blockage of size 2×2 is modeled by removing the corresponding vertices from the graph.

The paths for the solution shown in Fig. 3.1b are given by

$$r_1 = \Big((0, 4), (1, 4), (1, 3), (1, 3), (1, 3), (1, 3), (1, 3)\Big),$$
$$r_2 = \Big((2, 0), (1, 0), (0, 0), (0, 1), (0, 2), (0, 3), (1, 3)\Big), \text{ and}$$
$$r_3 = \Big((4, 1), (4, 2), (4, 3), (4, 4), (3, 4), (3, 4), (3, 4)\Big).$$

With this setup, it is possible to determine the complexity of the DMFB Routing Problem. This is the main result of this section and is formalized in the following theorem.

Theorem 3.1 *The DMFB Routing Problem is* NP-*complete.*

The proof of the Theorem is done via reduction of another, known-to-be NP-complete problem. The problem used is the $(n^2 - 1)$-Puzzle defined below.

Definition 3.3 $((n^2 - 1)$-**Puzzle, [RW90])** The aim of the $(n^2 - 1)$-Puzzle is to find a sequence of moves which will transfer a given initial configuration of an $n \times n$ board to a final (standard) configuration. A move consists of sliding a tile onto the empty square from an orthogonally adjacent square.

The question is: is there a solution for transforming the first (initial) configuration into the second (final) configuration requiring at most k moves?

Example 3.3 Consider the initial puzzle configuration in Fig. 3.2a. The goal is to reach the configuration in Fig. 3.2b in at most k steps. It turns out that the smallest of such k is 10.

As has been shown in [RW90], the $(n^2 - 1)$-Puzzle is one of the many NP-complete problems. Its structure already closely resembles the routing problem on digital microfluidic biochips. With the definition of the puzzle, the NP-completeness of the DMFB Routing Problem can now be easily proven.

Proof (Proof of the NP-*Completeness of the DMFB Routing Problem)* As commonly done in proofs for NP-completeness, see, for example, [GJ79], the proof is split into two parts. The first part proves that the problem lies within NP by showing that it is possible to guess a solution for the problem and verify that solution (or prove that it is, in fact, no solution) in polynomial time. The second part reduces a known NP-complete problem to the DMFB Routing Problem to show that it is at least as difficult as the reduced problem. Combining these parts concludes the proof that the DMFB Routing Problem is NP-complete.

The Droplet Routing Problem is in NP It is easy to guess a possible solution to the droplet routing problem. Algorithm 1 clearly verifies (or disproves) the solution in polynomial time. Assuming that the equality check can be performed in constant

1	8	2
	4	3
7	6	5

1	2	3
4	5	6
7	8	

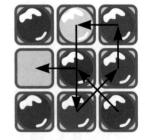

(a) Start configuration of the 8-puzzle

(b) End configuration of the 8-puzzle

(c) 8-Puzzle reduced to the DMFB Routing Problem

Fig. 3.2 Example of an 8-Puzzle. The configuration in (**a**) is to be transformed into the configuration shown in (**b**) by moving the numbered tiles. In (**c**), the problem reduced to the DMFB Routing Problem is shown. The arrows indicate the targets of the droplets representing the tiles

---

Algorithm 1: Verify guessed DMFB routing solution

Data: A DMFB Routing Problem $(\mathcal{D}, \mathcal{N}, T, G = (V, E), I)$
Data: A possible solution S of the droplet routing problem
Result: Decision whether S actually solves the problem

1 **for** $d \in \mathcal{D}$ **do**
2 $\quad$ **if** $r_d(1) \neq p_d^* \vee r_d(T) \neq p_d^\dagger$ **then**
$\quad\quad$ // Incorrect start and end points
3 $\quad\quad$ **return** *false*;
4 $\quad$ **for** $1 < t \leq T$ **do**
5 $\quad\quad$ **if** $\{r_d(t), r_d(t-1)\} \notin E$ **then**
$\quad\quad\quad$ // Invalid droplet movement
6 $\quad\quad\quad$ **return** *false*;
7 $\quad\quad$ **for** $d' \in \mathcal{D} \setminus n_d$ **do**
8 $\quad\quad\quad$ **if** $\{r_d(t), r_{d'}(t)\} \in I \vee \{r_d(t), r_{d'}(t-1)\} \in I$ **then**
$\quad\quad\quad\quad$ // Fluidic constraint violation
9 $\quad\quad\quad\quad$ **return** *false*;

10 **return** *true*

---

time and that membership testing is linear in the size of the set, no more than $\#\mathcal{D} \cdot (2 + T \cdot (\#E + 2 \cdot \#I \cdot \#\mathcal{D}))$ steps have to be performed.

Reduction of $(n^2 - 1)$ to the Droplet Routing Problem The reduction is straightforward. The board directly defines a quadratic biochip (see Example 3.2 for a similar biochip architecture) with

$$V = \{0, 1, \ldots, n-1\} \times \{0, 1, \ldots, n-1\}$$

and

$$E = \{\{(x_1, y_1), (x_2, y_2)\} \mid (x_1, y_1), (x_2, y_2) \in V \wedge |x_1 - x_2| + |y_1 - y_2| \leq 1\}.$$

The set I is chosen to contain the self-loops only. This means that the instance of the droplet routing only prevents multiple droplets on a single cell; no interference region around droplets is used. The dynamic fluidic constraints are still enforced, ensuring that only a single droplet moves in each time step. The tiles directly define the set of droplets; there is no multi-droplet net. That means that $\mathcal{D}$ and $\mathcal{N}$ are given by

$$\mathcal{D} = \{1, 2, \ldots, n^2 - 1\} \quad \text{and} \quad \mathcal{N} = \{((d, p_d^*), p_d^\dagger) \mid d \in \mathcal{D}\}.$$

The solution to the droplet routing problem gives $n^2 - 1$ routes that directly correspond to the solution of the $(n^2 - 1)$-Problem.

Example 3.4 The DMFB Routing Problem corresponding to the 8-Puzzle from Example 3.3 is shown in Fig. 3.2c. This reduction has been used to prove that the minimal value for k is 10.

One should note that the decision problem formulation does not directly solve the initial routing problem as it works on a fixed number of time steps. To actually use it to determine shortest routes, one needs to solve it repeatedly with an increasing T. This approach will be presented in detail in Sect. 3.4.

3.3 Heuristic Approaches

As has been shown in the previous section, the routing problem is inherently difficult. This is reflected in the fact that mainly heuristic approaches for solving the routing problem have been proposed so far.

An early work on routing on DMFBs uses the A^* algorithm to route droplets on DMFBs [Böh04]. In order to cope with the state space explosion, the droplets are assigned priorities. The droplets are then routed sequentially in order of descending priority. This means that higher prioritized droplets are routed first. For the routing problem, already routed droplets of higher priority are treated as mobile blockages while droplets of lower priority are ignored since they have not been routed and, therefore, do not introduce any blockages. This work employs the static fluidic constraints but may produce very long routes for the droplet routed at last. The paper does not use the concept of nets meaning that only independent droplets are routed.

The work [SHC06] does not only contribute the study of the fluidic constraints but also proposes a two-stage DMFB routing algorithm. The first step consists of determining M alternative routes for each net. All of these routes adhere to a timing constraint. In the second step, routes for each net are randomly chosen. This scheme prevents issues with droplet priorities which could lead to poor routes for the least prioritized droplets. This problem is called the net-routing-order dependence problem. The chosen routes are then evaluated by using the number of cells used in the overall routing as a cost function. Furthermore, the solution is checked for fluidic constraint violations. This process is repeated an adequate number of times until the set of routes with the minimum cost value is chosen.

In [CP08], the authors introduce the concepts of bypassibility for droplets and concession zones to which droplets can be routed in order to break up a deadlock. The main idea is to route the droplets in the order chosen by the bypassibility value. The non-routed droplets are then seen as blockages making the search space for the algorithm two-dimensional as no timing information is necessary. This work uses a slightly less restrictive version of the fluidic constraints, effectively putting only the horizontal and vertical neighbors of a cell in the interference region. This means that the interference region and the reachable positions are identical (meaning that in both cases the region as shown in Fig. 2.2a from Sect. 2.1.2 is used). In terms of the graph representation of the routing problem, this means that I and E are identical.

The BioRoute algorithm, proposed in [YYC08] divides the routing problem into two problems that are solved consecutively: global routing and detailed routing. Before performing any routing, the criticality for each net is computed. The criticality is a measure how difficult it is to route that specific net. In the global routing step, the approach is to iteratively search for a set of independent nets which then are routed using a network-flow-based algorithm order of decreasing criticality. This routing is performed on a "coarser" biochip which consists of cells representing a 3×3 array of cells on the original biochip. After all nets have been globally routed, the detailed routing part employs a negotiation-based algorithm that routes the droplets in decreasing order of criticality. This approach is not capable of directly handling 3-nets. It splits them into two 2-nets prior to routing.

The work [HH09] features an entropy-based algorithm for routing that makes use of preferred routing paths. The authors explicitly tackle the problem of droplet routing order by sorting the droplets based on the congestion of the routing regions. To model this, they borrow the notion of entropy from the field of thermodynamics, routing droplets with a higher variant in the entropy first. The main idea of this work is to mark rows and columns as preferred routing directions, penalizing droplets not directly following them. In a post-processing step, the routes are transformed into a one-dimensional representation and compacted using a dynamic programming approach. This work uses the same less restrictive version of the fluidic constraints as [CP08].

While all these heuristic approaches solve the routing problem, they cannot guarantee the minimality of the routes. This allows for relative comparisons between these approaches only. So far, it is not known how close to the technical optimum the solutions generated by these methods are. When transporting liquids that degrade over time, the minimality of routes can be of utmost importance.

Another aspect that is not addressed by these approaches is the correctness of the solutions. In classical circuit design, vendors validate their solutions using another tool from a different tool vendor to ensure that their netlists are indeed correct. While the presented methodologies most likely produce correct solutions, there is no guarantee.

3.4 Proposed Solution

The methodology proposed in this section (originally introduced in [KWD14]), in contrast to prior work, is non-heuristic and exact. The decision problem formulation from Sect. 3.2 is the main part of the proposed methodology. As already mentioned, a single decision problem is not sufficient for solving the routing problem. The general idea is to formalize the routing problem as a series of decision problems asking "Does there exist a routing in T time steps?" with an increasing T. This leads to the following, simple solving scheme:

1. Set $T = 1$.
2. Try to determine valid routes using at most T time steps.
3. If no such routes exist, increase T by one and go to 2.
4. Otherwise return routing solution.

Of course, if there is some a-priori knowledge about the minimal route length, the initialization of T can be adjusted accordingly.

The proposed approach has the following properties.

Correctness   The generated solutions are correct-by-construction with respect to the underlying model. This means that there is no need to further verify the solutions.

Minimality   As the iteration scheme starts with the minimal number of time steps and then iteratively increases it, finding the solution using the minimal number of time steps is guaranteed.

This guarantee is not just an important characteristic for concrete routing solutions but also allows to create *ground truth* for a variety of comparisons like the evaluation of heuristic approaches.

Abstraction   A formal model of the domain helps to understand the considered problems. The proposed approach is directly derived from the abstract model introduced in Sect. 2.2. This effectively frees the researcher from finding algorithms herself.

The proposed methodology inherently avoids problems like the net-routing order dependence problem (see Sect. 3.3) without resorting to workarounds such as splitting nets to work on 2-nets only.

Solving Time   As has been shown, the problem is NP-complete. This means that finding an optimal solution to the DMFB Routing Problem will take time. Instead of spending a lot of time, trying to find a good algorithm for solving the routing problem, a highly optimized solving engine is employed instead.

As the underlying technology to solve the decision problem, SAT (see Sect. 2.3) has been chosen. This means that the formal model is translated into a SAT instance that is then solved using an appropriate solver.

Creating a SAT instance is a two-step process. At first, the SAT variables used must be defined. In the second step, these variables are constrained to express states that adhere to the model only.

3.4.1   SAT Variables

To fully model the DMFB Routing Problem, Boolean variables that represent whether a certain droplet is present on a given cell position at a given time step are sufficient. That is, Boolean variables denoted

$$c_{p,d} \tag{3.3}$$

for $1 \le t \le T, d \in \mathcal{D}$, and $p \in \mathcal{P}$ are used. A truth value of $c_{p,d} = 1$ means that in time step t the droplet d is present at position p. The architecture of the biochip is implicitly modeled by the set of all positions $\mathcal{P}$ from which the indices p are taken. Blockages will be modeled using additional constraints, as will be shown in the next section.

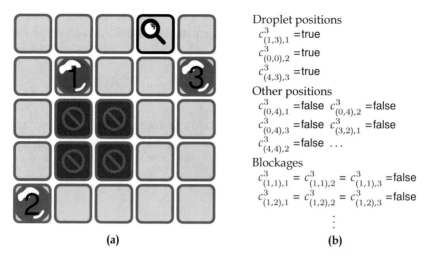

(a)   **(b)**

Fig. 3.3 Visualization of the third time step with exemplary variable assignment for the routing solution depicted in Fig. 3.1b. (**a**) Third time step of the routing solution. (**b**) Boolean variables and their assignments (excerpt)

How these variables are used to describe a configuration of a routing is illustrated in the following example.

Example 3.5 Consider again the routing problem and its solution from Fig. 3.1. The set of all cell positions is given by $\mathcal{P} = \{0, 1, 2, 3, 4\} \times \{0, 1, 2, 3, 4\}$, the set of droplets is given by $\mathcal{D} = \{1, 2, 3\}$ and the set of nets is given by

$$\mathcal{N} = \{(\{(1, (0, 4)), (2, (2, 0))\}, (1, 3)), (\{(3, (4, 1))\}, (3, 4))\}.$$

Figure 3.3a displays the third time step of the routing solution (the third time step means that the droplets have moved twice). The droplets' positions are given by $p_1 = (1, 3)$, $p_2 = (0, 0)$, and $p_3 = (4, 3)$. Figure 3.3b shows a representative subset of SAT variables as defined in (3.3) and their assignments. The blockage is realized by assigning all variables that correspond to the blocked positions the value false.

3.4.2 SAT Constraints

The SAT variables described above not only allow to describe valid biochip states but arbitrary ones. It is therefore necessary to constrain the way these variables can be assigned to add reasonable meaning to them.

In the following, constraints covering various aspects of routing are introduced and explained. The total of all these constraints then define the decision problem used in the iterative solving scheme.

Source and Target Configuration

In the first time step, the droplets are explicitly positioned on their target positions by directly setting the values of the corresponding SAT variables to **true** using the constraints

$$\bigwedge_{d \in \mathcal{D}} c^1_{p^*_d, d}. \tag{3.4}$$

To ensure that the droplets reach their target positions, the constraints

$$\bigwedge_{d \in \mathcal{D}} \left(\bigvee_{1 \leq t \leq T} c^t_{p^\dagger_d, d} \right) \tag{3.5}$$

are added to the SAT instance. Note that there is no specific time step in which the droplets are expected to be at their position. Only the fact that they eventually will is encoded. That means that droplets needing few time steps to reach their target may have left their target in the last time step.

Droplet Movement

A droplet may not arbitrarily appear on a cell on the biochip. It can only be present at a cell position p if it was already present in the neighborhood $N(p)$ of horizontally and vertically adjacent positions of that particular position p in the previous time step (see Fig. 2.2a for a visualization). This situation is depicted in Fig. 3.4. The corresponding constraints are

$$\bigwedge_{d \in \mathcal{D}} \bigwedge_{p \in \mathcal{P}} \bigwedge_{1 < t \leq T} \left(c_{p,d} \implies \bigvee_{p' \in N(p)} c^{t-1}_{p',d} \right). \tag{3.6}$$

The constraints model the movement of droplets starting with the second time step. This is necessary as the variable $c^{t-1}_{p',d}$ would be undefined for $t = 1$. The positions at the first time step are already fixed by (3.4).

The movement of the droplets is as unconstrained as possible in order to be as flexible as possible when determining a route. This, in turn, allows droplets to move around freely as long as they reach their targets in time, that is, before the droplet with the longest route reaches its target position. Even though this allows unnecessary droplet movement, it does not increase the number of time steps needed for routing.

Fig. 3.4 The movement of droplets is modeled by implications that go backwards in time by defining the valid positions in the previous time step. (**a**) Droplet position in time step t. (**b**) Valid droplet positions in time step $t-1$

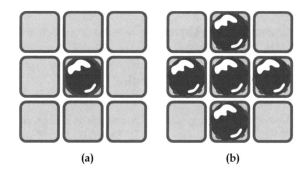

(a)   (b)

Fluidic Constraints

To avoid unintentional mixing, droplets d and d' that do not share a common net n must not enter each other's interference region. This means that d' must avoid the interference region $N_I(d)$ (see Fig. 2.2b for a visualization). This applies to the current time step as well as the previous one. This is formulated as follows:

$$\bigwedge_{n \in N} \bigwedge_{\substack{d \in \mathcal{D}_n \\ d' \in \mathcal{D} \setminus \mathcal{D}_n}} \bigwedge_{p \in \mathcal{P}} \bigwedge_{1 \leq \leq T} \left(c_{p,d} \implies \bigwedge_{p' \in N_I(p)} \neg c^t_{p',d'} \wedge \neg c^{t-1}_{p',d'} \right) \tag{3.7}$$

Note that in the case of $t = 1$ the variable $c^{t-1}_{p',d'}$ is undefined. In this case, it is simply removed from the constraint. For the sake of simplicity and readability, (3.7) does not explicitly depict this.

The common net sizes are 2-nets and 3-nets. Nevertheless, the presented routing method is technically capable of handling nets of arbitrary size.

Blockages and Consistency

To make the droplets respect the blockages on the biochip, the simple constraints

$$\bigwedge_{d \in \mathcal{D}} \bigwedge_{b \in \mathcal{B}} \bigwedge_{1 \leq \leq T} \neg c^t_{b,d} \tag{3.8}$$

are sufficient. The SAT variables for all droplets and time steps that correspond to blockages are directly set to **false**.

To ensure that a droplet is present at most once, the constraints

$$\bigwedge_{d \in \mathcal{D}} \bigwedge_{1 \leq \leq T} \left(\sum_{p \in \mathcal{P}} c_{p,d} \leq 1 \right) \tag{3.9}$$

are added to the SAT instance. These constraints do not enforce the presence of all droplets at every time step. This is in accordance with the constraints from (3.5) that do not fixate a droplet on its target once it has been reached. The movement constraints from (3.6) make sure that a droplet does not re-appear on the biochip.

3.5   Experimental Results

To evaluate the proposed methodology, its results are compared against the results of the approaches presented in Sect. 3.3. Experimental results have been generated for benchmarks commonly used in evaluating DMFB routing algorithms. The benchmark sets are described in [SC05] and [SHC06]. The benchmarks are organized in four sets containing multiple sub-problems. The in-vitro benchmarks describe a typical multiplexed experiment. The three human fluids samples urine, serum, and plasma are mixed with the reagents glucose oxidase and lactase oxidase and, afterwards, analyzed. This experiment is based on Trinder's reaction, for details see [Tri69]. In the protein experiments, samples with proteins are diluted. Afterwards, they are mixed with reagents for reaction. In the last step, the protein concentration is analyzed. For details on the experiment, see [Sri+04]. Figure 3.5 shows the second sub-problem of the in-vitro1 benchmark set. Table 3.1 summarizes the characteristics of the benchmarks.

The constraints of the previous section have been implemented in C++ using Z3 [DB08] as the SAT solver. The cardinality constraints of (3.9) were implemented using the scheme of [Sin05]. The experiments were run on a machine with four Intel Xeon CPUs at 3.50 GHz and 32 Gb RAM running Fedora 22.

Unfortunately, the proposed methodology cannot directly be compared against all approaches reviewed in Sect. 3.3. The work [SHC06] shows the applicability of the approach on a use case only. Furthermore, the results of [YYC08] that display the maximal route length as well as the average route length had to be taken from [HH09]. As the related work uses 0 as the first time step, their results have been adjusted to be comparable to the presented approach.

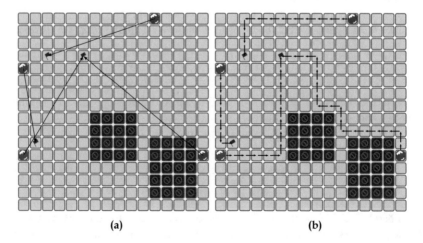

(a)   (b)

Fig. 3.5 Exemplary routing benchmark problem and one possible solution. (**a**) Sub-problem #2 of the in-vitro1 benchmark set. (**b**) Minimal solution to the problem using 19 time steps

Table 3.1 Characteristics of the benchmarks

Benchmark	Biochip size	# sub-problems	# nets	max d
in-vitro1	16×16	11	28	5
in-vitro2	14×14	15	35	6
protein1	21×21	64	181	6
protein2	13×13	78	178	6

#sub-problems: the number of sub-problems in the benchmark set, # nets: total number of nets in benchmark set, max d: maximal number of droplets in a sub-problem

Table 3.2 Comparison of proposed method with methods that use the fluidic constraints as defined in [SHC06]

Benchmark	BioRoute [YYC08]			Proposed			
	T	$\varnothing T$	#c	T	$\varnothing T$	#c	Dur (s)
in-vitro1	21	14.00	237	20	13.00	387	98.44
in-vitro2	18	12.33	236	17	11.07	391	62.39
protein1	21	17.31	1618	21	16.28	2318	1074.50
protein2	21	11.51	939	21	10.54	1455	172.24

T: maximal number of time steps needed, $\varnothing T$: average number of time steps needed, #c: number of used cells, Dur (s): time needed to solve the routing problems in seconds

Table 3.3 Comparison of proposed method with methods that use less restrictive fluidic constraints

Benchmark	High-performance [CP08]			Routability [HH09]			Proposed			
	T	$\varnothing T$	#c	T	$\varnothing T$	#c	T	$\varnothing T$	#c	dur (s)
in-vitro1	20	15.30	258	19	13.47	231	19	12.82	356	65.43
in-vitro2	21	13.00	246	18	11.43	229	17	11.07	379	49.13
protein1	21	17.55	1688	21	16.51	1588	21	16.28	2315	902.23
protein2	21	13.19	936	21	11.04	923	21	10.53	1456	132.67

T: maximal number of time steps needed, $\varnothing T$: average number of time steps needed, #c: number of used cells, Dur (s): time needed to solve the routing problems in seconds

No related work provides results for the individual sub-problems. Therefore, following the methodology of the other publications, the results are presented in aggregated manner. This means that for all four benchmark sets the maximal number of time steps needed to solve all sub-problems, the average number of time steps as well as the total number of used cells is reported.

As different authors use different versions of the fluidic constraints (see Sect. 3.3), (3.7) has been adapted accordingly. The difference is whether the diagonally adjacent positions are considered in the interference region. To incorporate this difference, the use of the neighborhood $N_I(p)$ in (3.7) is changed to $N(p)$. The comparisons to the related work are displayed in Table 3.2 for the fluidic constraints as defined in [SHC06] and Table 3.3 for the less restrictive fluidic constraints.

Table 3.4 Comparison of the relative overhead in cell usage

Benchmark	BioRoute [YYC08]	High-performance [CP08]	Routability [HH09]
in-vitro1	1.64	1.38	1.54
in-vitro2	1.66	1.54	1.66
protein1	1.43	1.37	1.46
protein2	1.55	1.56	1.58

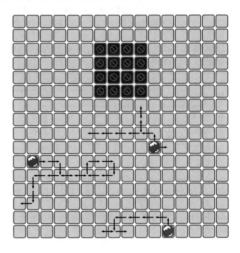

Fig. 3.6 Droplets taking detours in a routing solution. The figure shows a part of the exact solution of the third sub-problem of the in-vitro1 routing benchmark

The results clearly confirm that *exact* results can be determined in feasible run time for all benchmarks. This allows a qualitative comparison of previously obtained (heuristic) results to the actual minimum.

For all benchmarks, solutions with a lower average number of time steps could be generated. For the in-vitro benchmarks, the maximal number of time steps needed could be reduced as well. Nevertheless, since the differences are rather small, the proposed method is able to confirm that the previously proposed heuristics are already quite close to the actual optimum.

Comparing the results with respect to whether the fluidic constraints of [SHC06] or the less restrictive version was used shows that they have a rather insignificant impact on the length of the shortest possible routing solutions. Only for the *in-vitro1* benchmark set the maximal number of needed time steps is reduced by a single time step. The main improvement gained is the, relatively small, speed-up in solving time.

What can be seen is that the proposed approach uses more cells than the other solutions. The relative overhead in cell usage is shown in Table 3.4. Note that the proposed approach is in no way tailored to minimize the number of cells used. In fact, droplets that have no tight timing constraints can freely move around on the biochip (also see the remark in Sect. 3.4.2 and an exemplary visualization in Fig. 3.6). If the cell usage was to actually impose a hard criterion, simple post-processing techniques can be applied to heavily reduce the cell usage.

3.6 Summary

In this chapter, the routing problem for DMFBs has been formulated in a formal manner. This allowed to prove that the routing problem is NP-complete. Interestingly, the NP-completeness is caused by the dynamic fluidic constraints. Without the dynamic fluidic constraints, the routing problem might be deterministically solvable in polynomial time.

Due to the complexity of the problem, the formulation as a SAT instance and the use of a corresponding solver is a reasonable choice. The routing problem is exactly solved by iteratively solving the decision problem whether there exists a routing in T time steps. For each problem, T is increased until a solution is found. This guarantees that the exact optimal solution to the routing problem is found.

When compared to related work, it turns out that state-of-the-art heuristic approaches already produce close-to-optimal results. The approaches from the related work, additionally to determining short routes, try to optimize the number of cells used for routing. Even though the proposed method does not take the number of used cells into account at all, it never uses more than 1.7 times the number of cells as the solutions of the related work.

Chapter 4
Pin Assignment

4.1 Problem Formulation

In order to formulate the pin assignment problem, the concept of compatibility of actuation vectors is necessary.

Definition 4.1 (Compatibility of Actuation Vectors) Two actuation vectors $v, w \in \mathbb{A}^T$ are *compatible* (denoted as $v \circledast w$) if

$$(v_t = w_t) \vee (v_t = X) \vee (w_t = X)$$

holds for $1 \leq t \leq T$. In other words, two vectors are compatible if they either are identical for every time step or can be made identical by replacing a *don't care* with either 1 or 0.

Using the notion of compatibility, it is now possible to give the definition of a pin assignment and formulate the corresponding optimization problem.

Definition 4.2 (Pin Assignment) Let A be a set of actuation vectors of length T, that is, $A \subset \mathbb{A}^T$. A partition of A into k disjoint sets A_i is called *pin assignment of A* if, and only if, for every A_i the members of A_i are pairwise compatible, that is,

$$\forall v, w \in A_i : v \circledast w \qquad \text{for } 1 \leq i \leq k.$$

The cells/positions corresponding to the vectors of a set A_i are then assigned the same pin. The term p_v denotes the pin of a given vector v.

Example 4.1 Consider the actuation vectors

$$v_a = (0, 1, 0), \qquad v_b = (0, 0, 1), \qquad v_c = (X, X, 0),$$

© Springer International Publishing AG, part of Springer Nature 2019
O. Keszocze et al., *Exact Design of Digital Microfluidic Biochips*,
https://doi.org/10.1007/978-3-319-90936-3_4

$$v_d = (1, 0, 0), \qquad v_e = (0, 0, 0), \qquad v_f = (X, X, 0),$$
$$v_g = (0, 0, X), \qquad v_h = (0, 0, X), \text{ and} \qquad v_i = (X, X, X).$$

They correspond to the droplet movement of Example 2.4 from Sect. 2.2.3. The indices have been changed to allow for an easier reading. An excerpt of their compatibilities is given by

$$v_a \circledast v_c \qquad v_a \circledast v_f \qquad v_a \circledast v_i \qquad v_b \circledast v_g \qquad v_b \circledast v_h \qquad v_b \circledast v_i \qquad \ldots$$

One possible pin assignment would be

$$A_1 = \{v_a, v_c, v_f\} \qquad A_2 = \{v_g, v_e\} \qquad A_3 = \{v_h, v_b\} \qquad A_4 = \{v_d, v_i\}.$$

Using the pin sets to assign a pin to the vectors leads to

$$p_a = 1 \qquad p_b = 3 \qquad p_c = 1 \qquad p_d = 4 \qquad p_e = 2$$
$$p_f = 1 \qquad p_g = 2 \qquad p_h = 3 \qquad p_i = 4$$

Definition 4.3 (Pin Assignment Problem) Given a set of actuation vectors $A \subset \mathbb{A}^T$ and a fixed $k \in \mathbb{N}^+$, the *pin assignment problem* asks whether there exists a pin assignment of A using at most k sets. The problem is denoted by $\mathcal{PA}(A, k)$.

Note that in the above definitions, the actuation vectors were stored in a set. This implies that no vector occurs more than once. In real applications this is not necessarily the case (see Example 2.4 where the vector $(X, X, 0)$ appears twice). This problem is easily solved by removing any duplicates. One vector of these duplicates is kept as a representative of these vectors. The removed vectors will be assigned the pin of the representative. Furthermore, vectors consisting of X only can be removed completely as they have no influence on the solution with respect to k. Their pin can randomly be assigned or not be connected to the control logic at all. As duplicates impose technical problems only, in the following sets will be used. This improves readability and allows to focus on the core aspects. Any necessary pre-processing is performed implicitly.

4.2   Complexity of Pin Assignment

In this section, the complexity of the pin assignment problem is analyzed. Unfortunately, compatibility between actuation vectors is no equivalence relation. If it was, the pin assignment problem would be solvable in polynomial time. It is easy to see that compatibility is not transitive. Take the three actuation vectors $u = (0, 0)$, $v = (0, X)$, and $w = (0, 1)$. While u is compatible to v as well as v is compatible to w, it is easy to see that u is not compatible to w.

Different authors already noted that there is a close relationship between the pin assignment problem and other, NP-complete, problems such as *graph coloring* [XC06] or *partitioning into cliques* [XC08, YH14, DYH15]. But so far, no formal proof of the NP-completeness of pin assignment has been published.

The NP-completeness of the pin assignment problem will be shown by proving that it is equivalent to graph coloring. The proof is performed in two separate steps, one for each direction of the reduction, which are covered in dedicated subsections.

In order to be able to perform the actual NP-completeness proof, the graph coloring problem is reviewed first.

Definition 4.4 (Graph Coloring [GJ79]) Given a graph $G = (V, E)$ and a positive integer k with $k \leq \#V$, the graph coloring problems ask whether there exists a function

$$color : V \to \{1, 2, \ldots, k\}$$

such that $color(v) \neq color(v')$ whenever $\{v, v'\} \in E$. The graph coloring problem is denoted by $C(G, k)$.

The intuition for choosing this problem as the target for the reduction is that the colors will directly correspond to positions that can be assigned the same pin.

To show the equality between the problems, reductions that can be performed in polynomial time between C and $\mathcal{PA}$ need to be found.

4.2.1 Reduction from Pin Assignment to Graph Coloring

The first step is to construct the graph $G = (V, E)$ for C from the pin assignment problem $\mathcal{PA}(A, k)$. The vertices of G directly correspond to the actuation vectors. The set V, therefore, is identical to A. To construct the edges, a pairwise comparison of all vectors in A is performed. If two vectors a and b are incompatible, the edge $\{a, b\}$ is added to E. As $\circledast$ is symmetric, only $\frac{1}{2}n(n + 1)$ comparisons are needed, where n is the number of actuation vectors. The test for compatibility uses at most T comparisons. Adding all actuation vectors to the set of vertices takes n steps. In total, the reduction runs in $O(\frac{1}{2}n(n + 1)T + n)$. So the reduction to $C(G, k)$ can be done in polynomial time.

Example 4.2 Consider again the actuation vectors from Example 4.1.

The graph created by the reduction process of $\mathcal{PA}$ to C is shown in Fig. 4.1. The fact that $v_i = (X, X, X)$ is compatible to every other actuation vector is reflected in the fact that the corresponding node has no edges.

The next step is to prove that solving the constructed graph coloring problem actually does solve the initial pin assignment problem. Given the solution of $C(G, k)$, the solution of the initial $\mathcal{PA}(A, k)$ problem is constructed by generating the sets A_i as $A_i := \{v \in A \mid color(v) = i\}$ for $i = 1, 2, \ldots, k$. By definition of

Fig. 4.1 Graph constructed
in the reduction of pin
assignment to graph coloring

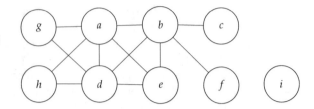

$C(G, k)$, all vertices are assigned exactly one color. The sets A_i form a partition of
A. The edges of E are not just separating neighboring vertices by color but, again
by construction, separate exactly those vectors that are incompatible. This leads to
the fact that within such an A_i indeed all vectors are compatible. This concludes the
proof that graph coloring solves the pin assignment problem.

4.2.2   Reduction from Graph Coloring to Pin Assignment

To construct the actuation vectors from the graph $G = (V, E)$, the adjacency matrix
$M = (m_{ij})$ of G is constructed. The rows of this matrix, after some alterations, will
serve as the actuation vectors for the pin assignment problem.

The zero entries $m_{ij} = 0$ in the matrix M encode that no connection between
the nodes v_i and v_j exists and that they, therefore, can have the same color. To
capture this choice in the pin assignment problem, all the zero entries are replaced
with don't care entries. The only exception for this is the main diagonal. Without
this exception, the vectors consist of 1's and X's only, making all vectors mutually
compatible. This would always allow a pin assignment using a single pin only. Now
the set A is constructed as $A = \{i\text{'th row of } M \mid 1 \leq i \leq n\}$.

Example 4.3 Consider the graph depicted in Fig. 4.1 that was created by the
reduction of $\mathcal{PA}$ to C. The corresponding adjacency matrix is shown in Fig. 4.2a.

Applying the transformation yields the matrix depicted in Fig. 4.2b which allows
to directly construct the following actuation vectors as input to the pin assignment
problem:

$$v_a = (0, 1, X, 1, 1, X, 1, 1, X) \qquad v_b = (1, 0, 1, 1, 1, 1, X, X, X)$$

$$v_c = (X, 1, 0, X, X, X, X, X, X) \qquad v_d = (1, 1, X, 0, 1, X, 1, X, X)$$

$$v_e = (1, 1, X, 1, 0, X, X, X, X) \qquad v_f = (X, 1, X, X, X, 0, X, X, X)$$

$$v_g = (1, X, X, 1, X, X, 0, X, X) \qquad v_h = (1, X, X, X, X, X, X, 0, X)$$

$$v_i = (X, X, X, X, X, X, X, X, 0)$$

	a	b	c	d	e	f	g	h	i
a	0	1	0	1	1	0	1	1	0
b	1	0	1	1	1	1	0	0	0
c	0	1	0	0	0	0	0	0	0
d	1	1	0	0	1	0	1	0	0
e	1	1	0	1	0	0	0	0	0
f	0	1	0	0	0	0	0	0	0
g	1	0	0	1	0	0	0	0	0
h	1	0	0	0	0	0	0	0	0
i	0	0	0	0	0	0	0	0	0

(a)

	0	1	X	1	1	X	1	1	X
v_a	0	1	X	1	1	X	1	1	X
v_b	1	0	1	1	1	1	X	X	X
v_c	X	1	0	X	X	X	X	X	X
v_d	1	1	X	0	1	X	1	X	X
v_e	1	1	X	1	0	X	X	X	X
v_f	X	1	X	X	X	0	X	X	X
v_g	1	X	X	1	X	X	0	X	X
v_h	1	X	X	X	X	X	X	0	X
v_i	X	X	X	X	X	X	X	X	0

(b)

Fig. 4.2 Adjacency matrix and its modification in the reduction of $C(G, k)$ to $\mathcal{PA}(A, k)$. (**a**) Adjacency matrix of the graph in Fig. 4.1. (**b**) Replacing the 0's by X but keeping the 0's on the main diagonal

An excerpt of their compatibilities is given by

$$v_a \circledast v_c \qquad v_a \circledast v_f \qquad v_a \circledast v_i \qquad v_b \circledast v_g \qquad v_b \circledast v_h \qquad v_b \circledast v_i \qquad \dots$$

One possible pin assignment is

$$p_a = 1 \qquad p_b = 2 \qquad p_c = 1 \qquad p_d = 3 \qquad p_e = 4$$
$$p_f = 1 \qquad p_g = 2 \qquad p_h = 2 \qquad p_i = 1.$$

The solution for this pin assignment problem also solves the original problem of Example 4.2 even though the actual actuation vectors are not identical. This is the desired behavior of the reductions.

To construct the rows of the altered adjacency matrix, at most (in the case of a complete graph) $\frac{1}{2}n(n-1)$ nodes need to be compared with each other, leading to $O((\frac{1}{2}n(n-1))^2) = O(\frac{1}{4}(n^4 - 2n^3 + n^2))$ steps. This shows that the reduction can be done in polynomial time.

It is left to prove that solving the $\mathcal{PA}$ instance actually solves the initial C problem. A partition of the vertices of G is induced by the vectors assigned the same pin (that is, the sets A_i) by solving the $\mathcal{PA}$ problem. So it is sufficient to show that the vertices corresponding to a pin induce a valid coloring of the graph G. This is true by construction as no two vectors can be assigned the same pin when there was an edge in the graph of the C problem as the 1 of the connection between the two vertices would need to merge with the 0 on the main diagonal of the other vertex.

4.3   Related Work

In the early work [SPF04], the authors present a biochip using less pins than cells available. They use a multiphase bus on linear pathways. This n-phase bus connects every nth electrode to the same pin, allowing droplets with a distance of $k \cdot n - 1$ for any $k \geq 1$ to be moved along the path. This addressing scheme is limited to parts of the biochip that form lines only. A line is a $1 \times n$ or $n \times 1$ region on the chip. The presented biochip uses direct addressing for the mixing region. This scheme requires a synchronization of the operations. It is not transferable to arbitrary rectangular biochips by simply employing the idea on each row (or column) individually as it would ignore the influence of neighboring cells on present droplets. The paper's focus is on the actual experiment and not too much on the technical aspects of the biochip itself. Interestingly, the presented biochip is comparatively small. From today's perspective, the need for pin reduction is not apparent.

In [HSC06], the authors come up with the idea of partitioning the biochip into separate regions. As long as there is no more than a single droplet in a region, five pins are sufficient to conduct arbitrary droplet movements. Therefore, a biochip with k regions only needs $5 \cdot k$ pins. Unfortunately, no automated scheme for determining the regions is provided. This issue is addressed in [XC06] where a two-phase approach is presented. In the first step, the regions are determined by analyzing the routes on which the droplets move during the experiment. The cells traveled by a droplet and their adjacent positions form a region. If there are overlaps between regions, these overlapping positions are used to form a new region. In the second step, for each region, the control pins have to be mapped to the electrodes. This is done by employing a strategy from the *Gomoku* board game.

The term *broadcast electrode-addressing* has been introduced in [XC08] where the pin assignment problem is tackled by reducing it to the clique partitioning problem. This NP-hard problem is solved by employing a heuristic algorithm.

In [HHC11], a progressive approach based on pin-count expansion is utilized. The idea in this paper is to create a reliability-oriented pin assignment meaning that electrodes should not be actuated unnecessary. By successively increasing the number of allowed pins and carefully choosing new subsets of non-addressed electrodes for which the pins are used, the authors create a pin assignment that prevents unnecessary electrode actuation.

In [LC13], the pin assignment problem is tackled at a more theoretical level. Lower and upper bounds for the number of needed pins on an $m \times n$ array are derived. Furthermore, an exact ILP-based algorithm is used to create ground truth for comparisons with heuristic approaches. The paper also presents one heuristic solution using a greedy approach.

The authors of [DYH15] present the idea of using a circuit with few inputs to drive the actual pins for the biochip. They also follow the clique partitioning approach, albeit in a slightly different manner. Instead of checking the compatibility between actuation vectors, they create status sequences by concatenating the actuation values at a time step. This procedure is performed for every time step, resulting in a set of vectors. Clique partitioning is then used to group these status sequences. This reduces the overall number of actuation states the biochip can assume. Then, for each electrode, a function to encode the electrode's state is constructed. Due to the reduced number of overall states, circuit sharing between these functions is possible, further reducing the required number of pins.

All these works have in common that they cannot guarantee to produce the optimal solution. Furthermore, it is currently not known how far from the optimum these approaches are. The main method proposed in this section deals with this issue by yielding the optimal solution with respect to the number of pins.

4.4 Proposed Solutions

In the following, two approaches for determining a pin assignment for given actuation vectors are presented. The first approach is a generic heuristic framework. It is optimized for speed and lacks the capability of guaranteeing to determine the optimal solution. The second solution is based on SMT and can guarantee the minimality of the solution with respect to the number of pins.

4.4.1 Heuristic Approach

In the following, a heuristic algorithm for creating a pin assignment for a given set of actuation vectors is presented. Due to the computational complexity of the problem, a heuristic approach is still valuable even though it cannot guarantee the minimality of the solution.

The idea behind the approach is to iteratively split the set of all actuation vectors into smaller sets that eventually will consist of compatible vectors only using a heuristic to deal with *don't care* values. The proposed approach actually provides a whole framework for solving the pin assignment problems as different heuristics with very different behavior can be chosen. The focus of this approach is on the execution speed.

The proposed approach is depicted in Algorithm 2. The framework starts with all vectors initially belonging to the same pin in line 1. Then for each time step all remaining sets will be split into two new sets, depending on the decision of the specified heuristic in lines 6–11. Afterwards, the set of sets that are used in the next time step is updated in line 12.

The heuristic framework cannot create solutions with a specified number of pins k or prove that no solution for k pins exists. That means that it does not solve $\mathcal{PA}(A, k)$.

This algorithm allows for a whole range of different heuristics to be used. The choice of a particular heuristic highly affects the number of pins used in the pin

---

Algorithm 2: Heuristic pin assignment

Data: Set $A = \{v_1, v_2, \ldots, v_n\}$ of n actuation vectors of length T
Data: A heuristic for assigning values to *don't care* values
Result: List of pin sets

```
1  S ← {A}
2  for 1 ≤ ≤ T do
3  │   S' ← ∅
4  │   for B ∈ S do
5  │   │   ones, zeros ← ∅
6  │   │   for v ∈ B do
7  │   │   │   if heuristic X assignment = 1 then
8  │   │   │   │   ones ← ones ∪ {v}
9  │   │   │   else
10 │   │   │   │   zeros ← zeros ∪ {v}
11 │   │   S' ← S' ∪ {ones, zeros}
12 │   S ← S'
13 return S
```

---

assignment. Among the many possible heuristics, the following five choices will be used in this chapter.

constant   Replace all X values by a fixed value. Both, replacing all X' by 0's and replacing all X's by 1's, have been tested.

alternate   Alternately replace the X values by 0 or 1.

even out   Replace the X value in such a manner that the sets *zeros* and *ones* are even in size. The heuristic is parameterized in the decision which set to fill when both are of equal size.

maximize   Replace the X value in such a manner that one of the sets *zeros* and *ones* grows faster. The heuristic is parameterized in the decision which set to fill when both are of equal size.

random   Randomly replace X by either 0 or 1.

Example 4.4 Consider again the actuation vectors from Example 2.4. As a heuristic, all *don't care* values are replaced by a 0. The corresponding steps of the algorithm are depicted in Table 4.1. The columns *ones* and *zeros* show the split that occurred in the step. The ordering of the sets, separated by a semicolon, corresponds to the ordering of the sets within S.

The algorithm finds a solution using four pins. This solution is, by chance, minimal. The corresponding pin sets are $A_1 = \{v_a\}$, $A_2 = \{v_b\}$, $A_3 = \{v_d\}$, and $A_4 = \{v_c, v_e, v_f, v_g, v_h, v_i\}$.

The result of this heuristic framework is very dependent on the heuristic. Already choosing to replace every X with 1 instead of 0 leads to a solution using six pins.

Note that the actual implementation does check whether the process can be terminated. It does not make sense to actually add empty sets or try to split sets

Table 4.1 Finding a pin assignment for the actuation vectors from Example 2.4 using the heuristic algorithm and the heuristic that alternatingly substitutes *don't care* values with 1 and 0

t	B	Ones	Zeros
1	$\{v_a, v_b, v_c, v_d, v_e, v_f, v_g, v_h, v_i\}$	$\{v_d\}$	$\{v_a, v_b, v_c, v_d, v_e, v_f, v_g, v_h, v_i\}$
2	$\{v_d\}$	$\emptyset$	$\{v_d\}$
	$\{v_a, v_b, v_c, v_e, v_f, v_g, v_h, v_i\}$	$\{v_a\}$	$\{v_b, v_c, v_e, v_f, v_g, v_h, v_i\}$
3	$\{v_d\}$	$\emptyset$	$\{v_d\}$
	$\{v_a\}$	$\emptyset$	$\{v_a\}$
	$\{v_b, v_c, v_e, v_f, v_g, v_h, v_i\}$	$\{v_b\}$	$\{v_c, v_e, v_f, v_g, v_h, v_i\}$

with single elements, as done in the previous example. These checks are not shown in Algorithm 2 for educational purposes in order to not obscure the main idea behind the approach.

4.4.2 Exact Solution

As pin assignment is an NP-complete problem, using an SMT solver is a promising method for achieving solutions that use as few pins as possible. For this, the usual encoding for the graph coloring problem is chosen (see, for example, [Knu15]). Instead of manually encoding the colors, which serve as the pin numbers, SMT is used to work directly on integer numbers.

The encoding states what cells must not be assigned the same color. This is an exclusive encoding in the sense that only explicitly forbidden choices are modeled.

The instance consists of natural number variables pin_v that are created for every vector $v \in A \subset \mathbb{A}^T$ with additional constraints that enforce the following restrictions:

1. Incompatible vectors must not be assigned the same pin.
2. The number of pins has an upper bound of k.

The first restriction is realized by testing each pair of actuation vectors for compatibility. If two incompatible vectors v and w are found, the corresponding variables pin_v and pin_w must not be equal, that is,

$$\bigwedge_{\substack{v,w \in A \\ \neg(v \circledast w)}} pin_v \neq pin_w \qquad (4.1)$$

is added to the SMT instance.

For the second restriction, let k denote the maximal number of pins to use. The following simple constraints ensure that no more than k pins are used:

$$\bigwedge_{v \in A} pin_v < k \qquad (4.2)$$

This allows to check whether $\mathcal{PA}(A, k)$ holds. If an assignment using k pins exists, the solver also returns it (the values of the pin_v variables contain the pin number).

Additionally, to determine the smallest k for which an assignment exists, the same idea as with determining the minimal number of time steps needed for routing, as done in Sect. 3.4, is employed. The process of determining the minimal pin assignments begins with setting k to 1 (or a known lower bound). If no such satisfying assignment exists, the value of k is increased by 1 and the instance is, once again, given to the SMT solver. This process is iterated until a solution is found. Note that there always exists a solution for the pin assignment for $k = \#A$ as this corresponds to the case of a *directly addressing biochip*. This ensures that the iterative process will successfully terminate.

When looking at Example 2.4, it can be seen that there is a vector, namely v_i, that can be assigned with any pin. In order to speed up the solving process, all such vectors are removed from the problem instance. As there always has to be at least one pin, these vectors can safely be assigned to this pin.

Example 4.5 Consider again the actuation vectors of Example 2.4. The corresponding SMT instance for $k = 4$ is given by

$$pin\_a \neq pin\_b \wedge pin\_a \neq pin\_d \wedge pin\_a \neq pin\_e \wedge pin\_a \neq pin\_g \wedge pin\_a$$
$$\neq pin\_h \wedge pin\_b \neq pin\_c \wedge pin\_b \neq pin\_e \wedge pin\_b \neq pin\_d \wedge pin\_b$$
$$\neq pin\_f \wedge pin\_d \neq pin\_e \wedge pin\_d \neq pin\_g \wedge pin\_d \neq pin\_h$$
$$\wedge pin_a < 4 \wedge pin_b < 4 \wedge pin_c < 4 \wedge pin_d < 4 \wedge pin_e < 4$$
$$\wedge pin_f < 4 \wedge pin_g < 4 \wedge pin_h < 4.$$

The inequalities can directly be taken from Fig. 4.1.

Note that only the constraints that compare the values of the *pin* variables to the upper bound of pins need to be changed when looking for a solution using more pins. This can directly be exploited when using solvers that support partial removal of assertions. If an instance is not satisfiable, only the smaller-than assertions have to be replaced by updated versions. This allows the solver to reuse any information it has gained about the inequality assertions in the previous solving step.

4.5 Experimental Results

In this section, the pin assignment solutions will be analyzed. Firstly, both the heuristic and exact method are evaluated on actuation vectors for routing solutions. In the second step, the exact pin assignment approach is used to optimize the existing pin assignment of a commercially available biochip.

4.5.1 Evaluation of the Pin Assignment

4.5.1.1 Used Benchmarks

To generate input data for both the heuristic framework and the exact approach, the optimal routings determined by the exact routing approach from the previous chapter are taken. From these routings, the corresponding actuation vectors were extracted (see Example 2.4 for details). These vectors serve as a good basis for evaluating the solutions against each other. These vectors do not allow to compare the proposed solutions against related work, as they do not correspond to the vectors used in the other works. As, unfortunately, the routings from the related approaches are not available, no direct comparison is possible.

As already discussed in Sect. 3.4.2, the routes of the solutions determined by the exact, SAT-based approach may contain a lot of detours. Such detours (see Fig. 3.6 for an example) significantly increase the number of necessary pins as well as the complexity of the problem instance. For the sub-problems *in-vitro1_3, in-vitro2_3, in-vitro2_8, protein1_30, protein2_37,* and *protein2_46*, after a timeout of 60 min still no solution was determined and the solving process terminated. These sub-problems therefore are removed from the benchmarks considered in the rest of the section.

4.5.1.2 Heuristic Approach

To evaluate the heuristic approach, five different simple heuristics for replacing the *don't care* values were selected: *constant, alternate, even out, maximize*, and *random* (see Sect. 4.4.1 for a detailed explanation).

Neither of these heuristics is too sophisticated as the aim is to use heuristics that are easy to understand and describe. Furthermore, the computational complexity should be low in order to generate results as fast as possible. That means that in the inherent speed/quality trade-off that comes with every heuristic approach, the decision is to completely choose speed.

The results for applying these heuristics in the framework are shown in Table 4.2 in rows 2 till 10. The columns "max k" show the maximal number of pins needed in one of the sub-problems; the columns "$\varnothing k$" show the average number of pins needed over all sub-problems. Results for all benchmark sets and all heuristics are computed less than 2 s in total so no run-time is reported. Only for the constant *don't care* replacement, the parametrization has a significant influence on the number of pins. Interestingly, the concrete pin assignments do vary even in cases where the maximal number of pins as well as the average pin set size are identical.

The similar results of the *alternate* and *even* heuristics are easily explained. When starting with empty sets *zeros* and *ones* (see Algorithm 2), evening them out is close to alternatingly choosing 1 or 0. The small differences result from actuation vector entries already having a defined value of 1 or 0 that are put into one of the sets without the help of the heuristic.

Table 4.2 Summary of all presented methods for pin assignment ordered by the number of used pins

Method	in-vitro1		in-vitro2		protein1		protein2	
	max k	$\varnothing k$	max k	$\varnothing k$	max k	$\varnothing k$	max k	$\varnothing k$
Exact	13	7.7	8	5.9	14	7.6	12	6.5
Run-time (s)	181.6		3.5		129.7		43.9	
Const 0	30	16.4	24	12.5	41	20.6	35	13.5
Maximize 1	83	35.7	46	23.9	109	45.4	71	24.4
Maximize 0	83	35.7	46	23.9	109	45.4	71	24.4
Const 1	123	57.6	79	37.4	158	66.8	108	35.6
Rand	149.0	75.7	107.9	52.2	243.9	99.2	130.5	45.6
Std. dev.	1.7	1	3.6	1.2	2.6	0.6	2.7	1.2
Alternate 1	149	76.9	108	54.2	244	100.1	131	47.3
Alternate 0	149	76.9	108	54.2	244	100.1	131	47.3
Even 0	149	77.1	109	54.8	244	100.0	131	48.1
Even 1	149	77.5	114	55.5	244	100.4	131	48.9

For the exact approach, additionally the run-time is shown; for the random heuristic, additionally the standard deviation is presented

Always choosing 0 as a replacement for the *don't care* is, by far, the best heuristic from the set of evaluated heuristics.

To evaluate the behavior of the random heuristic, the pin assignment was performed 10,000 times and the mean result and the standard deviation were computed. This is done in order to avoid an inaccurate analysis due to outliers. The results are presented in rows 7 and 8 of Table 4.2. As a uniform distribution for the generation of random numbers was used, the results obtained are similar to the ones for the *alternate* and *even* heuristics. The results are, on average, slightly better. The standard deviations show that these results usually vary by at most three pins, except for the in-vitro2 case where the standard deviation is larger than three (but still less than four). Choosing a different distribution for the random numbers will shift the results towards the *constant* heuristic.

Overall, choosing the *constant* heuristic with a choice of 0 is the best choice.

4.5.1.3 Exact Solution

The pin assignment results for the exact pin assignment method are summarized in the first row of Table 4.2. One can see that, except for the benchmarks removed from the sets, the solving time is quite short. These values allow to judge the results from the heuristic approach.

One can see that the difference in the number of necessary pins between the exact solution and the heuristic results is rather large.

4.5.1.4   Comparison of the Heuristic and Exact Results

Compared to the ground truth generated by the exact solutions, the proposed heuristic method generates very poor results. The concrete heuristics that were used were chosen to be simple in order to be easy to understand and fast in computation. Both of these goals were reached.

Table 4.2 gives an overview over all presented methods by ordering them in descending order of quality with respect to the number of pins used.

4.5.2 Optimizing the Pin Assignment

Pin assignment is not just applicable when minimizing the pin count for a given protocol. It can also be applied to analyze and optimize existing pin assignments. Given an existing biochip, such as the commercial one presented in [LC13] where the pin assignment is done by hand, the question is, whether a pin assignment using less pins is possible. Applying exact algorithms to find a pin assignment gives insight about the optimality of the biochip.

The procedure for this is to concatenate the actuation vectors of movements that should be performable on the chip. If the choice of actuation vectors is representative for the usual operation of the biochip, the resulting pin assignment can then be used for normal biochip operation.

In this section, the *reaction region* and *routing region* from the biochip depicted in Fig. 4.3 are evaluated. For both regions, a representative building block and representative nets are chosen (they are taken from [ZC12] where no nets for the detection region were provided). In [ZC12], the problem is solved by applying pin-aware routing, which will be covered in detail in Sect. 5.1. These regions

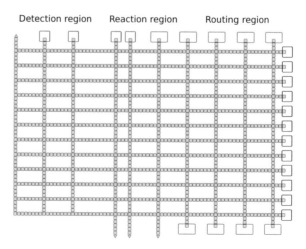

Fig. 4.3 Commercially available biochip manufactured by [ALL] (picture reproduced from [ZC12, Figure 5])

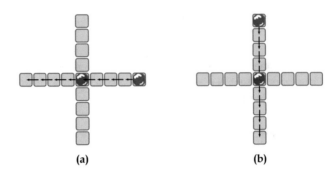

Fig. 4.4 Representative movements in the routing region. (**a**) Droplets moving from $(4, 4)$ to $(0, 4)$ and from $(8, 4)$ to $(4, 4)$. (**b**) Droplets moving from $(4, 8)$ to $(4, 4)$ and from $(4, 4)$ to $(4, 0)$

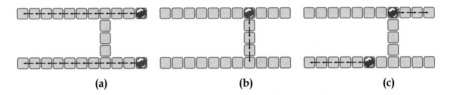

Fig. 4.5 Representative movements in the reaction region. (**a**) Droplets moving from $(10, 4)$ to $(0, 4)$ and from $(10, 0)$ to $(0, 0)$. (**b**) Droplet moving from $(7, 4)$ to $(7, 0)$. (**c**) Droplets moving from $(5, 0)$ to $(0, 0)$ and from $(7, 4)$ to $(10, 4)$

of the biochip only allow for uniquely determined shortest routes: the direct movement of droplets towards their target position. This allows to directly extract the corresponding actuation vectors (also see Example 2.4) without having to solve the pin-aware routing problem. Figures 4.4 and 4.5 show the considered routes.

The results of the different pin assignment methods are compared in Table 4.3. Interestingly, all methods use only six pins for the routing region. This leads to the impression that all approaches behave identical or at least similar as they all use the proven least number of pins. In this region, also the pin assignment chosen by the manufacturer already is close to optimal. In the reaction region, the optimal assignment using six pins is less than 50% of the manufactured pin assignment using 14 pins. One can see that in cases where the situation becomes more complicated (more diagonal influence of electrodes, longer actuation vectors) the non-exact methods need more pins than necessary whereas the exact solution is still capable of determining the exact minimum. Note that the proposed results for the reaction region are better than the one of the ILP-based method of [ZC12]. This is probably due to an upper bound on the computing time spent for solving the problem exactly.

The exact pin assignments, as determined by the proposed method, are shown in Fig. 4.6. In contrast to the results from [ZC12], the exact method does not introduce new pins for the vertical paths on for the reaction region. For positions $(7, 1)$ and $(7, 3)$, pins number 0 and 1 are re-used while the ILP solution from [ZC12] introduces the new pins 8 and 10.

Table 4.3 Comparison of the number of control pins generated by different approaches

	Biochip [ALL]	ILP [ZC12]	Heuristic [ZC12]	Heuristic [LC13]	Proposed method
Routing region	7	6	6	6	6
Reaction region	14	8	10	14	6
Total	21	14	16	20	12

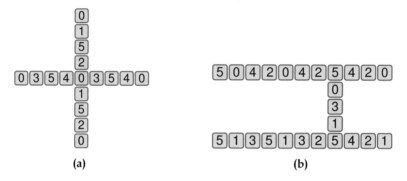

(a)   (b)

Fig. 4.6 Pin assignments for the two building blocks. (**a**) Minimal pin assignment for the movements in the routing region (see Fig. 4.4). (**b**) Minimal pin assignment for the movements in the reaction region (see Fig. 4.5)

4.6   Summary

In this chapter, the NP-completeness of the pin assignment problem has been proven. While NP-completeness has been previously conjectured in the literature, no formal proof was given so far.

Two methods to solve the pin assignment problem have been presented: one is a heuristic framework that allows for the use of different heuristics and the other one is an exact solution guaranteeing the minimality of the solution. The NP-completeness of the problem motivates the use of a reasoning engine to determine the exact solutions.

In many cases, the use of an SMT solver is feasible for determining a solution to the pin assignment problem. In contrast to the exact results, the heuristic framework produces rather poor results. This loss in optimality comes with a drastic reduction in run-time. It remains an open task to find more suitable heuristics that, by sacrificing run-time, will yield better results.

This chapter also showed that problems that seem to have a very narrow scope, in this case assigning the minimal number of pins to cells, can be used to solve a more general problem. This more general problem is to analyze and optimize a commercially available DMFB. This problem could be successfully solved by iteratively solving correspondingly adapted pin assignment problems.

Chapter 5
Pin-Aware Routing and Extensions

5.1 Pin-Aware Routing

Determining routes and a corresponding pin assignment simultaneously is different to the problems discussed in the previous two chapters: there are two criteria for optimization—the length of the routes and the number of pins. This combined problem is called *pin-aware routing problem*.

As the pin-aware routing problem solves both the routing problem and the pin assignment problem, the following theorem trivially holds.

Theorem 5.1 *Pin-aware routing is* NP-*complete*.

This, again, justifies the use of an SMT solver for determining the solutions. In the following, a corresponding method for this is presented (originally proposed in [Kes+15]). More precisely, the next section introduces the SMT formulation for it. Then, after a short discussion of the related work, four different use cases are discussed.

5.1.1 SMT Formulation

The combined SMT formulation for routing and pin assignment can almost be done by simply combining both previous SAT/SMT formulations. The following three different types of variables are used to create the SMT formulation.

- *Droplets* The Boolean variables $c_{p,d}$ indicate whether droplet d is present at position p in time step t.
- *Pins* The natural number variables pin_p model the pin number for the electrode at position p.

© Springer International Publishing AG, part of Springer Nature 2019   55
O. Keszocze et al., *Exact Design of Digital Microfluidic Biochips*,
https://doi.org/10.1007/978-3-319-90936-3_5

- *Electrode Actuation* The Boolean variables act_p^t indicate whether the electrode at position p is actuated in time step t. For a given position p, the variables $(act_p^1, \ldots, act_p^T)$ form the actuation vector of p.

The SMT formulation does not only determine routes and a pin assignment that allows to perform that routing, but also the actuation vectors that are necessary to actually move the droplets as desired.

The only issue when combining the previous formulations is that the information about which cells must not share the same pin is not given a-priori. To cope with this, the actuation vectors act_p are part of the solution space. That means that the solver fills them while determining a solution. These vectors are then used to generate the constraints necessary for pin assignment.

That two electrodes must not share the same pin is determined as follows:

$$\bigwedge_{1 \leq \leq T} \bigwedge_{p, p' \in \mathcal{P}} act_p^t \neq act_{p'}^t \Rightarrow pin_p \neq pin_{p'}$$

as opposed to by the a-priori given graph structure from (4.1).

The actuation states of the variables act_p^t are determined as follows. Whenever a droplet occupies a cell, the corresponding electrode has to be actuated. This is modeled by the constraints

$$\bigwedge_{p \in \mathcal{P}} \bigwedge_{d \in \mathcal{D}} \bigwedge_{1 \leq \leq T} c_{p,d} \Rightarrow act_p^t.$$

It needs to be ensured that no undesirable droplet movements happen when the determined actuations are used to move the droplets on a physical biochip. Without further constraints, arbitrary electrodes, besides those necessary for the intended droplet movement, may be actuated. Therefore, the following constraints ensure that

- no electrode next to the droplet is actuated and that
- exactly one electrode is actuated in the following time step.

The first point makes sure that the droplet is on a well-defined position whereas the second point prevents uncontrolled movements. The corresponding constraints are given by

$$\bigwedge_{p \in \mathcal{P}} \bigwedge_{d \in \mathcal{D}} \bigwedge_{1 \leq \leq T} c_{p,d} \Rightarrow \left(\bigwedge_{\substack{p' \in N_I(p) \\ p' \neq p}} \neg act_{p'}^t \wedge \sum_{p' \in N_I(p)} act_{p'}^{t+1} = 1 \right).$$

The constraints for the droplet movements are identical to the ones from the previous chapter.

5.1.2 Related Work

Some routing algorithms discussed before already incorporate pin assignment; see Sects. 3.3 and 4.3 for the corresponding discussions.

The work [HH10] begins with formulating an exact ILP instance for pin-aware routing. In order to reduce the solving time, the solution is then split into two sub-problems. The first sub-problem determines global routing tracks using the A^* algorithm. The second sub-problem is the actual routing along the tracks. For this, droplets are routed using the ILP formulation in an iterative manner. This means that droplet routes are determined one after another; each route restricting the next routes.

In [ZC12], the authors solve the routing and pin assignment problem in a combined fashion. The means to increase the solving speed are to split a routing problem into multiple sub-problems. The authors use their methodology to investigate the pin assignment of commercially available biochips and show that many pins can be removed without losing the generality of the chip.

For all of these works, the same argument from Sect. 3.3 applies: The minimality and correctness is not inherently guaranteed for heuristic approaches. Therefore, an exact methodology can provide valuable ground truth. Please note that exact solutions for sub-problems do not imply that a solution consisting of solutions for these sub-problems is also an exact solution for the whole problem.

5.1.3 Use Cases

In this section, four different use cases for pin-aware routing are presented. These use cases illustrate the versatility of the SMT formulation of Sect. 5.1.1. The main idea to tackle these use cases is always the same: fix the aspect of the routing that is known by explicitly enforcing values for the corresponding variables and let the solver fill in the values for the variables that represent the problem at hand.

5.1.3.1 Routing with Pin Assignment

The most general case that can be addressed by the proposed method is to perform the routing and pin assignment task in a combined fashion. For this, no values for the routes or the pins are known. This problem formulation is the most unconstrained one from the four considered use cases.

As stated above, both the number of time steps and the number of pins can be optimized. For this, many different schemes can be used. The following example, which is taken from the literature, minimizes the number of pins.

Example 5.1 Consider the routing problem in Fig. 5.1a, which is taken from [HH10]. Three droplets are to be routed and the corresponding pin assignment

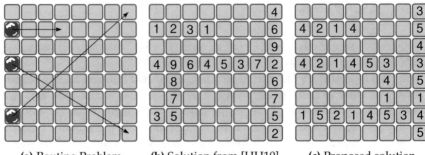

(a) Routing Problem   (b) Solution from [HH10]   (c) Proposed solution

Fig. 5.1 The three droplets in (**a**) are to be routed using as few pins as possible. (**b**) and (**c**) display the pin assignment of cells used in the routing process. Cells not used for routing are left blank

is to be performed. In this example, the upper bound for the number of time steps is 14 and the number of pins is the criterion that is to be optimized. This fixed number of time steps is chosen as it is used in the related work. Also, 14 is the minimal number of time steps allowing to find routes for the given problem.

The result from [HH10] uses nine pins in total (see Fig. 5.1b). The proposed method is able to prove that only five pins are required to move all the droplets to their target positions. For this solution, droplet routes differing from the solution from [HH10] are determined (see Fig. 5.1c).

5.1.3.2 Routing Following Given Pins

Commercially available biochips often are designed to solve a specific task. Figure 5.2 shows the layout of a biochip by [ALL] that is used for n-plex immunoassay. The pin assignment for this chip is determined by the manufacturer and cannot be changed. The manufactured pin assignments for the routing region of the biochip are presented in Fig. 5.3.

To perform routing on such a chip, therefore, the fixed pins need to be taken into account. The presented exact pin-aware routing method is easily capable of doing so. The variables pin_p simply need to be explicitly set to the value $assigned_p$ chosen by the manufacturer. This is realized by adding the following constraints to the SMT instance.

$$\bigwedge_{p \in \mathcal{P}} pin_p = assigned_p$$

Any routes that are now determined by the proposed approach will be realizable using the given pin assignments.

Fig. 5.2 Layout of a
commercial biochip used for
n-plex immunoassay
manufactured by [ALL]
(picture adapted
from [HHC11, Figure 12])

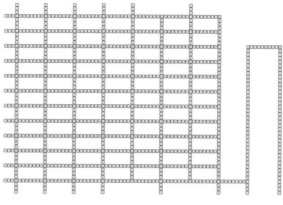

Fig. 5.3 Pin assignment of
the routing region (picture
adapted from [HHC11,
Figure 13])

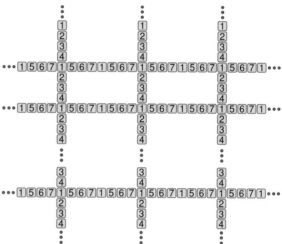

5.1.3.3   Check for Routability

As already noted in [HSC06], some pin assignments prevent certain movements.
Such assignments are undesirable as they reduce the applicability of the biochip for
different protocols and, therefore, can be considered as a design flaw.

Example 5.2 Consider the situation inspired by [HSC06, Figure 4(a)] depicted in
Fig. 5.4. The droplet is to be moved from the lower right corner to the left-most
column. Again, the method to solve this problem is to explicitly set the values of the
pin_p variables in the SMT formulation. As can be seen easily, it turns out that no
routing solution for this problem exists. In contrast to existing heuristic solutions,
the proposed method is able to actually prove that no solution for a given number of
time steps exists.

Fig. 5.4 Routing problem
with a given pin assignment.
No routing is possible. The
droplet can only advance to
the second column from the
left without being split

This can be used to test the biochip's pin assignment before actually manufactur-
ing it, saving time and money. The procedure for this is as follows:

1. Define a set of nets that should be routable on the given biochip and
2. define an upper bound for the number of time steps for each routing problem.

This test set is then evaluated using the proposed method. If each net can be routed,
the biochip conforms to the specification. Obviously, the choice of the problems
that are used in the test set determines the overall quality of the evaluation. Note
that heuristic methods inherently are not capable of proving the absence of routings
and the faultiness of the chip.

This method can also be used to verify a routing solution from another, possibly
not pin-aware, algorithm. In this case, both the $c_{p,d}$ and act_p variables have known
values. The answer of the SMT solver then determines whether the solution is valid
on the chip with that particular pin assignment.

5.1.3.4 Optimizing the Pin Assignment

The original work that optimized the given pin assignment of a commercially
biochip did so by solving the routing and pin assignment problem at the same
time [HSC06]. In Sect. 4.5.2, the exact pin assignment solution was used to
determine an optimized pin assignment. To generate the input for that approach,
in a separate step, the exact routing solution was used to create droplet routes and
the corresponding actuation vectors. Using the pin-aware routing, there is no need
to split the process into two separate steps anymore. As expected, the pin-aware
routing yields, up to renaming of the pins, the same results as in Sect. 4.5.2.

5.2 Routing with Timing Information

So far no routing algorithm is capable of dealing with droplets having multiple
routing targets or (dis)appearing on the biochip at some point in time. The first
situation, for example, arises when a droplet should be routed to a waste reservoir
after having been routed to an optical detector. The second situation, for example,
occurs when a droplet is dispensed to the biochip or is the result of a splitting
operation. These situations could not be dealt with using the existing routing

solutions. Therefore, the benchmarks were divided into sets of sub-problems at such droplet splits. These sub-problems can then be solved individually.

Even when solving these sub-problems exactly, there is no guarantee that the initial problem is solved in the optimal manner. The proposed exact routing solution is, after minor adaptation, capable of doing exactly this. For doing so, droplets and blockages have to be enriched with timing information (this has originally been proposed in [KWD14]).

The blockages in the routing problems represent fluidic operations currently being performed. As these operations will eventually finish, every blockage $b \in \mathcal{B}$ is assigned a start and end time, $_b^*$ and $_b^\dagger$, respectively. The blockages constraints from Eq. (3.8) are changed to

$$\bigwedge_{d \in \mathcal{D}} \bigwedge_{b \in \mathcal{B}} \bigwedge_{_b^* \leq t \leq _b^\dagger} \neg c_{p,d}.$$

This extension is already sufficient to analyze interesting aspects of temporal routing. The blockages present in the benchmarks are most likely over-approximating their presence. Not all fluidic operations that are blocking parts of the biochip may take the whole time to finish. Therefore, considering blockages as permanent is not necessarily modeling the biochip's behavior accurately.

Figure 5.5 shows a situation where a permanent blockage leads to a route that uses seven time steps in total. If the blockage was gone in time step 4, the total routing time would decrease.

To evaluate the influence of temporal blockages, four routing problems are investigated using different timings for the blockages. For this purpose, the routing problems introduced in [HH09, Fig. 1 and 2] and the sub-problem #59 of the *protein2* benchmark set have been used as they feature many blockages.

As baseline results to compare against, results from the exact routing method are used. They provide the minimal number of time steps T needed to route when the blockages are permanent.

Then, these routing problems are solved again with blockages being present in 90%, 70%, and 50% of the time steps. The intervals in which the droplets

Fig. 5.5 Simple routing situation: a permanent blockage leads to a solution using seven time steps

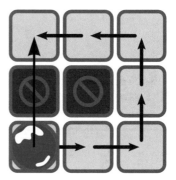

Table 5.1 Exploitation of temporary blockages

Benchmark	Size	Initial T	90%		70%		50%	
			T	$\varnothing T$	T	$\varnothing T$	T	$\varnothing T$
Protein2 #59	13×13	17	17	17	5	12.1	5	6
[HH09, Fig. 1]	13×13	16	16	16	16	16	14	15.4
[HH09, Fig. 2 (a)]	13×13	14	14	14	14	14	11	12.3
[HH09, Fig. 2 (b)]	16×16	25	25	25	24	24.0	22	23.7

are present are generated randomly. For each benchmark and each blockage, 10 different instances of the routing problem are generated.

The results are provided in Table 5.1. The initially determined number of time steps (initial T) denotes the best possible result currently available. Then, for each benchmark and each length of intervals, 10 instances are considered. The best (*min*) and average ($\varnothing$) number of time steps T are reported in the remaining columns. The results show that routing can indeed be significantly improved if blockages are not assumed to be present the entire time.

In the best case, the number of time steps is reduced from 17 to 5. This demonstrates the potential of using timing information of blockages.

To overcome the limitation of having to solve the individual steps in an experiment using individual routing sub-problems, timing information for the droplets needs to be added. As was done with the blockages, every droplet d is assigned its spawn time $\overset{*}{_d}$ and removal time $\overset{\dagger}{_d}$.

The source and target configuration from Eqs. (3.4) and (3.5) for the droplets becomes

$$\bigwedge_{d \in \mathcal{D}} c^{d}_{p^{*}_{d}, d} \wedge c^{d}_{p^{*}_{d}, d}. \tag{5.1}$$

Note that in this setting, the precise time step in which the droplet arrives is important. Also, the consistency constraints from Eq. (3.9) need to take into account the spawning and removal of droplets. This is done by splitting them into three parts.

The constraint

$$\bigwedge_{d \in \mathcal{D}} \bigwedge_{\overset{*}{_d} \leq t \leq \overset{\dagger}{_d}} \left(\sum_{p \in \mathcal{P}} c_{p,d} = 1 \right)$$

ensures that the droplet is present exactly once during the course of the experiment. The constraints

$$\bigwedge_{d \in \mathcal{D}} \bigwedge_{1 \leq t < \overset{*}{_d}} \neg \left(\bigvee_{p \in \mathcal{P}} c_{p,d} \right)$$

and

$$\bigwedge_{d \in \mathcal{D}} \bigwedge_{\overset{\dagger}{d} < t \leq T} \neg \left(\bigvee_{p \in \mathcal{P}} c_{p,d} \right)$$

prevent droplets from being on the biochip outside their specified time interval.

Consider again the pin optimization problem for a given pin assignment from Sect. 4.5.2. In Sect. 5.1.3.4 it has already been solved in a more convenient way. The routing and pin assignment step could be performed simultaneously. Still, multiple consecutive pin-aware routing problems had to be solved. Taking into account the temporal aspects as described above, the pin optimization of a given biochip can now be performed in a single step.

In principle, the benchmark sets could also be solved in one single step. This would not only guarantee the optimality on all the sub-problems but also guarantee the optimality of the overall solution for the experiment. Unfortunately, the sub-problems of the benchmarks do not seamlessly fit together. Technically, it would be possible to solve them but from an experimental point of view it would not make much sense as there would be "jumps" between the individual parts.

Starting with an experiment description, determining a concrete biochip behavior would be synthesis, a topic that is dealt with in the next chapter. Consequently, the benchmark sets are not solved in a single instance.

5.3 Aging-Aware Routing

As already noted in [HH09], for reasons of reliability, the use of electrodes on a biochip should be limited. The issue of electrode breakdown is extensively studied in the literature, see, for example, [Hu+13, XC09].

One way to address this issue is to minimize the overall number of cells used in the experiment. Related work, see, for example, [CP08, YYC08, HH09], therefore considers this in their approaches. Again, this can easily be incorporated in the existing exact pin-aware routing solution by adding a constraint. Let u be the threshold for the overall cell usage. The corresponding constraint, ensuring that not too many cells are used, is given by

$$\sum_{p \in \mathcal{P}} \sum_{1 \leq t \leq T} \bigvee_{d \in \mathcal{D}} c_{p,d} \leq u. \qquad (5.2)$$

The idea is to simply sum up the usage of the cells over all positions and time steps. The $\bigvee_{d \in \mathcal{D}}$ checks whether the cell at position p is used by any droplet in time step t.

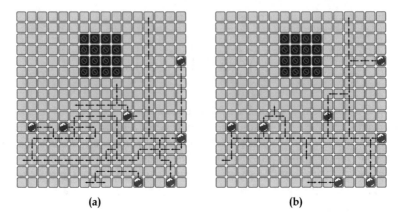

(a)   (b)

Fig. 5.6 Solutions for sub-problem #3 of the in-vitro1 benchmark sets using different amounts of cells. (**a**) Routing solution using 88 cells with at most five usages of an individual cell. (**b**) Routing solution using 54 cells with at most 12 usages of an individual cell

Example 5.3 Recall the solution for the third sub-problem of the in-vitro1 benchmark set depicted in Fig. 3.6. The full solution uses 88 different cells. The maximal number of cell usages for an individual cell is 5. This corresponding solution is shown in Fig. 5.6a.

The solution using the minimal number of cells using the minimal number of 19 time steps is shown in Fig. 5.6b. The number of detours is reduced at the cost of additional individual cell usage due to droplets waiting on the routes. The maximum number of usages of an individual cell is 12.

As can be seen in the exact results above, the number of cells used is reduced at the cost of the usage of individual cells. The assumption that the reduction of overall cell usage will prevent or delay breakdown, therefore, is questionable. It seems more reasonable to limit the maximal usage on the individual cells.

The actuation restriction on cells can easily be done by introducing an individual threshold u_p for each position p. This threshold is then enforced by adding the constraint

$$\sum_{t=1}^{T} \bigvee_{d \in \mathcal{D}} c_{p,d} \leq u_p \qquad (5.3)$$

for all positions p.

These constraints can either be used on their own or can be combined with the overall usage constraint from (5.2).

The thresholds u_p may vary from electrode to electrode. This allows to use a-priori knowledge about the biochip and its individual cells. On a biochip that is used for the first time, they are assumed to be identical.

Example 5.4 Consider again the third sub-problem of the in-vitro1 benchmark set. Instead of minimizing the total number of cells that are used, as has been done in Example 5.3, the individual actuations of cells are to be restricted using (5.3).

Figure 5.7 displays routing solutions for different threshold values u_p of 5, 4, 3, and 2. No solution for using cells only once is possible as there are two nets that need to cross each other. Note that the solution for $u_p = 5$ differs from the solution shown in Example 5.3. The solution in the other examples uses 88 cells in total whereas the current solution uses 78 cells only. This is due to the fact that the solver is given a problem that now restricts the individual cell usage. As the problem differs, the solver may choose different heuristics when solving. The overall number of cells used is not constrained in any way allowing this degree of freedom.

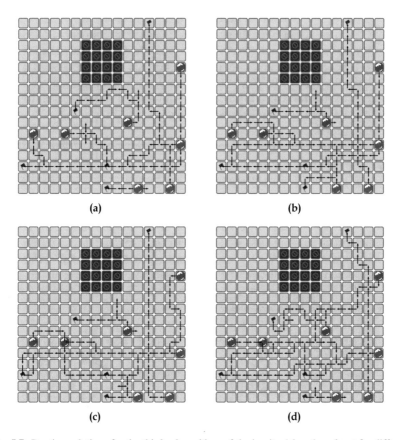

Fig. 5.7 Routing solutions for the third sub-problem of the in-vitro1 benchmark set for different threshold values u_p. (**a**) Routing solution with $u_p = 5$ using 78 cells. (**b**) Routing solution with $u_p = 4$ using 83 cells. (**c**) Routing solution with $u_p = 3$ using 82 cells. (**d**) Routing solution with $u_p = 2$ using 87 cells

As expected, a smaller value of u_p, limiting the number of actuations of individual cells leads to an increase in the overall cell usage. The degrees of freedom the solver has allows for the case of $u_p = 4$ with an overall cell usage of 83 and the case of $u_p = 3$ using 82 individual cells; one cell less than the former case. Nevertheless, the overall trend is clear.

5.4   Routing with Different Cell Forms

So far, only cells of square shape have been considered in this work. These are, by no means, the only possible shapes cells can take. Recently, triangularly shaped cells have been proposed in [Dat+14]. Here, the authors argue that cells of equilateral triangular shape are superior to quadratic shapes; especially when performing a mixing operation. Earlier, hexagonally shaped cells were introduced in [SC06b] where they are expected to increase the effectiveness of droplet transportation. Also, the shape is beneficial in the case of physical electrode breakdowns when droplets have to be re-routed. In [Fré+02], less regular cell shapes were investigated.

Biochips are not restrained to having cells of identical shape only. In [RBC10], a biochip specialized on sample preparation is presented. This biochip uses a rotary mixer built from skewed cells.

It is an open question which cell shapes are most suitable for performing droplet routing. This section (based on [Sch+17]) analyzes the routing behavior for different cell shapes. The routability is measured by comparing the time steps needed to route on the given cell shape in comparison to the exact lower bound for square shaped cells provided by the exact routing solution from Chap. 3.

Only cell shapes that form a tessellation of the droplet movement area are considered. Biochips using these shapes can perform the most general experiments. This limits the choice of cell shapes to triangles, squares, and hexagons. Figure 5.8 shows pictures of the cell shapes under investigation as fabricated for a real-world DMFB. Biochips, such as presented in [RBC10], are specialized for certain experiments and will, therefore, not be considered.

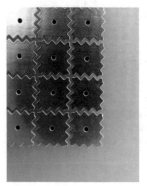  

Fig. 5.8   Different cell shapes for a digital microfluidic biochip

5.4.1   Problem Formulation

In the following, the impact of the cell shapes on the routability will be investigated.

The hypothesis is that the fluidic constraints are the main limiting factor when routing droplets. The assumption is that the impact of the fluidic constraints can be inferred from the ratio between the fields that are reachable from a cell with a given shape and the fields that need to be kept free to enforce the fluidic constraints.

Figure 5.9 displays the reachable cells and the interference regions for the different cell shapes. Originating from the center droplet, the arrows indicate the reachable cells. All shown cells are part of the interference region. The only exception is the relaxed assumption from [Dat+14] for triangles, where only the reachable cells are part of the interference region. These situations lead to the ratio shown in Table 5.2.

Depending on whether the conservative interference region of 12 cells or the interference region from [Dat+14] is used, the expected routability order of the shapes is as follows:

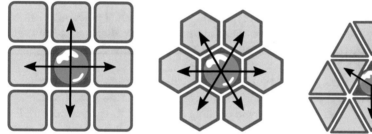

Fig. 5.9 Reachable cells and cells in the interference region of the droplet in the center for the three considered cell shapes. Note that for the triangular shape, the authors of [Dat+14] only consider the reachable cells to be part of the interference region

Table 5.2 Interference region to reachable fields ratios for the investigated cell shapes

Shape	Interference Region	Reachable fields	Ratio
△	12/3	3	$\frac{1}{4}/1$
▢	8	4	$\frac{1}{2}$
⬡	6	6	1

For the triangular shape both the conservative fluidic constraints and the less restrictive fluidic constraints from [Dat+14] are reported

conservative assumption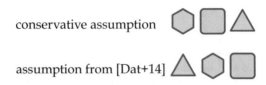

assumption from [Dat+14]

To check the hypothesis, the routing benchmarks used throughout this book are taken, transformed to use the different cell shapes and then routed using the exact routing algorithm from Chap. 3. The results are then compared with respect to the number of time steps needed to move all droplets to their targets.

5.4.2 Transformation of Routing Problems

Due to the lack of benchmarks for biochips using either hexagonal or triangular cells, the benchmark sets in-vitro1, in-vitro2, protein1, and protein2 are transformed to problem instances using these shapes. This process is straightforward and is illustrated on the routing problem shown in Fig. 5.10a.

To create a hexagon biochip, every second row of the initial problem is shifted to the left by half a cell. Cells that are adjacent in this new alignment of cells are the ones that are connected when replaced by a hexagonal cell. The resulting grid is depicted in Fig. 5.10b.

To transform a row of square cells to triangle cells, the first cell of the first line is replaced by a triangle pointing downwards (the origin is at the lower left corner). The next cell is then replaced by a triangle pointing upwards. This is repeated until the whole row is using triangles. The whole process for the row is iterated for all rows. For each row, the starting orientation (downwards/upwards) of the first triangle is switched. The resulting biochip is shown in Fig. 5.10c.

One can see that the two cells right of the blockage form a separated region on the biochip using triangular shapes. This can lead to unsolvable routing problems when the start or end positions of a net are located in such a region. All of the 168 routing problems were investigated for separated regions. Benchmarks containing such regions are removed from the benchmark sets. This resulted in a total of 136 routing problems suitable for all three cell shapes.

5.4.3 SMT Formulation

For investigating the effects of the different cell shapes the SMT formulation does not need to be changed or extended. The formulation already is stated in such a general way that only the input data needs to incorporate the cell shape under investigation. This can easily be done by adjusting $c^1_{p^*_d,d}$, $c^1_{p^\dagger_d,d}$, $N(p)$, $N_I(p)$ and $\mathcal{B}$ accordingly (see the previous section).

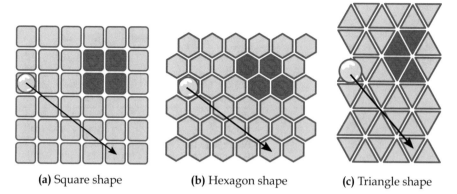

(a) Square shape   **(b)** Hexagon shape   **(c)** Triangle shape

Fig. 5.10 (**a**) Initial routing problem. (**b**) Routing problem transformed to use hexagon shaped cells. (**c**) Routing problem transformed to use triangle shaped cells

5.4.4   Experimental Results

The experimental evaluation investigated the routability of cell shapes for the four configurations: square shaped cells, hexagonal shaped cells, and triangular shaped cells with a conservative assumption on the interference region as well as the assumption from [Dat+14]. Overall, 136 benchmarks were evaluated.

The routing results for the square shaped cells serve as the baseline that all other results are compared against.

The results are shown in Table 5.3. The lines "in-vitro1," "in-vitro2," "protein1," and "protein2" show the necessary time steps for routing on the benchmark sets. The next four lines depict results on the aggregated benchmarks relative to the aggregated values of the results for the square shaped biochips.

The results for the hexagonal shapes are in line with the hypothesis. On average, roughly two time steps could be saved when using hexagons resulting in an overall decrease in needed time steps of 15%.

The results for the triangular shaped cells use more time steps than the routes for the square shaped cells. The overhead is approximately 40% of the solution for square shaped cells. While the increase in time steps is what the hypothesis predicted, interestingly the size of the interference region does not have a significant influence. Therefore, the hypothesis could be rephrased to state that the number of reachable cells is the limiting factor when routing a droplet.

The results clearly show that, even under the assumptions from [Dat+14], with respect to routability, triangular shapes are the least favorable choice of cell shape. The hexagonal cells perform slightly better than the square shapes.

Table 5.3 Benchmark results for routing on squares, hexagons, and triangles with and without the conservative interference region assumption

Shape	⬛	⬡	△	△
Interference region size	8	6	12	3
in-vitro 1	127	103	180	179
in-vitro 2	166	138	202	199
protein 1	1022	885	1434	1432
protein 2	456	384	679	673
Total steps necessary	1768	1510	2495	2483
Total Difference	0	-258	+727	+715
Average Difference	0	-1.9	+5.4	+5.3
Percentual Change	0%	-15%	+41%	+40%

5.5 Routing for Micro-Electrode-Dot-Array Biochips

So far, routing related problems in the domain of "conventional" DMFBs have been considered. Recently, a new technology, referred to as *micro-electrode-dot-array* (MEDA), has been proposed [WTF11, Wan+14, LYL15, Lai+15]. For this technology, routing still is the main challenge. In the following, an exact routing approach for MEDA biochips (originally proposed in [Kes+17]) will be presented that borrows its ideas from the previous, exact routing methodologies.

5.5.1 Motivation and Background

Digital microfluidic biochips are a technology that has matured over the years. However, today's DMFBs still suffer from several limitations. These limitations include

1. the constraint on the fine-grained control of droplet sizes and volume,
2. the lack of integrated sensors for real-time detection, and
3. the reliability/yield concern of fabricated DMFBs.

In order to overcome these limitations, a new technology, the MEDA technology, has been proposed. Unlike conventional DMFBs, the MEDA architecture is based on the concept of a sea-of-micro-electrodes with an array of identical basic microfluidic unit components called *microelectrode cells* [WTF11]. Such a cell consists of a microelectrode, a control circuit, and a sensing circuit [Lai+15]. The microelectrode cell provides fine-grained control of droplet-manipulation operations and real-time sensing/actuation for each microelectrode in an independent manner. Prototypes of MEDA biochips have been fabricated using TMSC 0.35 μm technology. In these devices, a power-supply voltage of 3.3 V is used for the controlling and sensing circuits [LYL15, Lai+15].

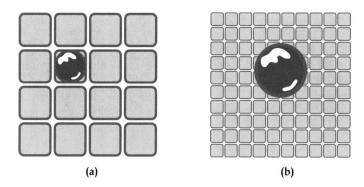

Fig. 5.11 Droplet-Cell correspondence for classical and MEDA DMFBs. (**a**) Classical DMFB: one-to-one correspondence between droplets and cells. (**b**) MEDA DMFB: a droplet may cover more than a single cell

In this technology, microelectrodes can dynamically be grouped together to form functional electrodes or fluidic units. These units are, for example, mixers and diluters. This grouping provides better reconfigurability and programmability compared to conventional DMFBs. Droplets and electrodes are in no one-to-one correspondence any more (see Fig. 5.11). A droplet covers more than a single electrode. This allows for droplets to move diagonally and even change their shapes.

However, due to the inherent differences between conventional DMFBs and MEDA biochips, existing droplet routing techniques cannot be utilized for MEDA biochips. Some of these differences include:

- MEDA biochips allow for a precise and flexible control of the droplet size and droplet shape.
- When manipulating droplets with different shapes, the corresponding actuation and resistive forces may differ. These differences result in variable droplet velocities, which also have to be considered during droplet routing.
- Finally, MEDA biochips allow to diagonally route droplets. These diagonal movements provide a higher degree of freedom compared to conventional DMFBs where droplets can only be moved horizontally or vertically. This may be exploited for shorter routings.

Two aspects, diagonal movement and shape change, are visualized in Fig. 5.12.

The problem solved in this section is the routing problem for MEDA-biochips. Despite the differences in the droplets' possible behavior, many similarities with the "conventional" routing problem remain. The fluidic constraints, for example, still must not be violated. The previously presented solution for exact droplet routing still gives valuable insight and inspiration for the proposed solution in this section.

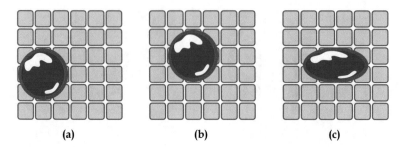

Fig. 5.12 Possible droplet movements on MEDA biochips: (**a**)–(**b**) diagonal movement, (**b**)–(**c**) droplet shape change

5.5.2 MEDA Model and Problem Formulation

A MEDA biochip consists of a rectangular-organized sea-of-micro-electrodes, which can be represented as a grid of size $W \times H$, where W is the width and H is the height. Each of these electrodes has a unique (x, y)-position, where $0 \le x < W$ and $0 \le y < H$.

An experiment executed on a MEDA biochip usually consists of multiple droplets. The set of all droplets, again, is denoted by $\mathcal{D}$. Even though droplets on MEDA biochips can assume complex shapes, in the following, only rectangular shapes are used. Rectangular in this context means that the droplet covers cells that form a rectangle, meaning that there are no "arms." The rationale behind is that it becomes more difficult to move droplets in non-rectangular shapes with respect to cell actuation as well as to keep that particular shape. The rectangular shape of a droplet can be described by specifying its *aspect ratio*. An aspect ratio of $a : b$ specifies a rectangle with a width of a cells and a height of b cells.

The movement of droplets is achieved by activating micro-electrodes which produce a force moving the droplets. In [Li+16], a simplified *velocity model* describes the dynamic response induced by the forces acting on a single droplet [BG06, OVB12]. The only actuation force is electrowetting-on-dielectric, and the resistive forces include plate shear force, viscous drag force, and contact-line friction force. The velocity model of [Li+16] considers the droplet as a single mass. Furthermore, the droplets are assumed to move with an average velocity and the velocities for droplets can be calculated using their size and shape. Therefore, we define the velocity on a per-droplet basis, that is, each droplet $d \in \mathcal{D}$ is associated with a specific velocity. The velocity is represented as the number of time steps required by a droplet to move to a neighboring position on the grid. This means that a single time step represents the time that the fastest droplet needs to move to a neighboring position. Slower droplets need multiple time steps to move to a neighboring position on the grid. Formally, each droplet $d \in \mathcal{D}$ is associated with a *movement modifier* m_d, which specifies the time steps a droplet stays on the same position before reaching a new location on the grid.

Another unique characteristic of MEDA biochips is its *morphing property* which means that the droplet aspect ratio can be changed. This offers new advantages for droplet routing by allowing a higher degree of freedom. For example, this allows droplets to form a thin line in order to move through a small corridor between two blockages. However, due to the fact that actuation forces cannot always overcome resistive forces [BYM07], droplets cannot be morphed arbitrarily. To represent the allowed droplet aspect ratios an *aspect ratio library* R_d is introduced for each droplet $d \in \mathcal{D}$.

To describe the location and aspect ratio of a droplet, a pair of positions on the grid is used. The first position specifies the lower left corner and the second position specifies the upper right corner of the droplet. In the following, this is written as $((x^\downarrow, y^\downarrow), (x^\uparrow, y^\uparrow))$. The droplet aspect ratio $a : b$ can then be computed via $a = x^\uparrow - x^\downarrow + 1$ and $b = y^\uparrow - y^\downarrow + 1$.

As with conventional biochips, some electrodes of the grid may be occupied by *blockages*, for example due to on-going fluidic operations, and, therefore, cannot be used for droplet routing. The set of blockages is denoted by $\mathcal{B}$. Similar to droplets, the locations and sizes of these blockages are also defined by pairs of positions. With the notion of *location* instead of *position* the concept of source and target positions p_d^* and $p_d^\dagger$ as well as *nets* remain the same as in Chap. 2.

Example 5.5 To denote that a droplet with a droplet aspect ratio of $3 : 2$ is located three electrodes from the left side and on the top of a 10×6 grid, the notation $p = ((2, 4), (4, 5))$ is used. Similarly, a $2 : 2$ blockage at with its lower left corner at position $(4, 1)$ is denoted by $b = ((4, 1), (5, 2))$.

This situation is visualized in Fig. 5.13. The two droplets are to be routed to the common target location $((7, 3), (9, 4))$, indicated by the dashed rectangle. As the aspect ratio of the target position is $3 : 2$, droplet 2 needs to change its shape in order to be successfully routed. Droplet 1 can be routed without a change in the aspect ratio.

Fig. 5.13 Two droplets of droplet aspect ratio of 3:2 and 2:2, respectively, are to be routed to the region highlighted by the dashed rectangle

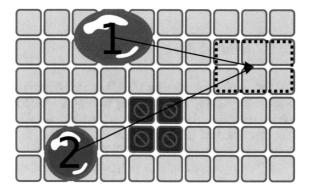

Given these prerequisites and terminology, the MEDA routing problem can then be defined as follows.

Definition 5.1 (MEDA Routing Problem) The input of the problem consists of

- the biochip architecture, given by the width W and height H of the biochip,
- the, possibly empty, set $\mathcal{B}$ of blockages,
- an optional droplet aspect ratio library R_d for every droplet d,
- an optional movement modifier m_d for every droplet d, and
- the set $\mathcal{N}$ of nets.

The MEDA routing problem is to determine routes for all nets that do not violate the fluidic constraints and respect all the blockages present on the biochip as well as the movement modifiers and the aspect ratio libraries. As a second problem, the minimization of route lengths can be considered.

The proposed method of this section is again to model the routing problem using SMT. Again, SMT instances are iteratively solved until a solution is found. This approach is justified as the routing problem for MEDA biochips is inherently difficult. When the ability to move droplets in diagonal directions is not used, the following theorem holds.

Theorem 5.2 *The routing problem for MEDA biochips without diagonal movement is* NP-*complete.*

Proof By fixing the droplet size to 1×1 and restricting the movement speed of all droplets to one cell in one time step, the problem is identical to the routing problem of Chap. 3 and solves the $(n^2 - 1)$-puzzle.

So far, no proof for the NP-completeness of the routing problem for MEDA biochips allowing diagonal movement is known allowing to formulate the following conjecture only.

Conjecture 5.1 The routing problem for MEDA biochips is NP-complete.

5.5.3 Related Work

Existing routing solutions for DMFBs are not directly applicable to MEDA biochips because the routing algorithm has to consider the MEDA specific characteristics mentioned above. Therefore, there is a need for specific routing solutions for MEDA biochips. The two research papers [Li+16, Che+11] are the only previous works considering droplet routing for MEDA biochips so far.

In [Li+16] a technique is proposed to approximate the droplet routing time. This however basically relies on the distance between the start and target positions and takes neither the actual routing paths nor dependencies between droplets into account. In [Che+11], an A^*-based approach is proposed to tackle the droplet routing problem for MEDA biochips. The algorithm is able to route multiple droplets with different sizes and supports diagonal movements of droplets. However, this approach employs a heuristic for which the difference in the number of time steps compared to the optimal solution is not yet known.

Overall, no solution exists which is capable of determining *exact* results for the routing problem of MEDA biochips. This issue is addressed with the solution proposed in the next section.

5.5.4   Proposed Exact Routing Approach

As has been done in the previous chapters, the general idea is to take the MEDA model and transform it into a symbolic formulation that models all possible paths of droplets including their velocities and the morphing property for a specified number of time steps. This formulation is then passed to an SMT solver, which is used to determine a satisfying assignment representing the routing of the droplets. If such an assignment can be determined, the obtained values to the variables of the symbolic formulation represent an explicit routing of all droplets. If no such assignment exists (proven by the SMT solver), the number of considered time steps is increased until a satisfying solution is determined. This way, as long as one exists, the determination of an exact solution is guaranteed.

5.5.4.1   SMT Formulation

As in the previous exact routing solutions, variables for droplets, combined with appropriate constraints, are sufficient to model the routing process. The main difference is that in this case, variables from the domain of the natural numbers are used. This is necessary as droplets and cells are not in a one-to-one relationship any more.

Definition 5.2 (Symbolic Representation of Droplets) Consider a routing problem with droplets $d \in \mathcal{D}$ and a maximal number of time steps T. Then, a *symbolic representation* of all possible solutions can be defined over natural number valued variables

$$x_d^{\downarrow}, y_d^{\downarrow}, x_d^{\uparrow}, y_d^{\uparrow}$$

which are created for every droplet $d \in \mathcal{D}$ and the number of considered time steps $1 \leq t \leq T$.

These variables describe the position and the droplet aspect ratio of droplet d at time step t. We use the following abbreviation to make the following formulas easier to read

$$p_d^t = \left((x_d^{\downarrow}, y_d^{\downarrow}), (x_d^{\uparrow}, y_d^{\uparrow}) \right).$$

Example 5.6 Consider again the routing problem from Example 5.5 and assume $T = 5$. Assigning the id 1 to the upper droplet and the id 2 to the lower droplet, the following variables provide a symbolic formulation of all possible routings:

$$x_1^{\downarrow,1}, y_1^{\downarrow,1}, x_1^{\uparrow,1}, y_1^{\uparrow,1}, x_1^{\downarrow,2}, \ldots x_1^{\downarrow,5}, y_1^{\downarrow,5}, x_1^{\uparrow,5}, y_1^{\uparrow,5}$$

$$x_2^{\downarrow,1}, y_2^{\downarrow,1}, x_2^{\uparrow,1}, y_2^{\uparrow,1}, x_2^{\downarrow,2}, \ldots x_2^{\downarrow,5}, y_2^{\downarrow,5}, x_2^{\uparrow,5}, y_2^{\uparrow,5}.$$

For example, setting $x_1^{\downarrow,1} = 2$, $y_1^{\downarrow,1} = 4$, $x_1^{\downarrow,2} = 2$, $y_1^{\downarrow,2} = 3$ represents a movement of the lower left corner of droplet 1 down one cell between time steps 1 and 2.

By allowing arbitrary assignments to these variables, all possible routings of the droplets within the considered time steps are symbolically represented. An unconstrained set of variables obviously allows for arbitrary routings—including illegal configurations like overlapping droplets. In order to avoid that, semantic meaning has to be attached to the variables by adding further constraints. Therefore, the following constraints are added that model the various aspects of routing on MEDA biochips.

Restricting Droplets to the Grid

The most important and most obvious constraint ensures that the droplets stay within the boundaries of the biochip. This means that the values of the corresponding SMT variables must be bounded by W and H. The constraint ensuring that droplets may occupy valid grid cells only is given by

$$\bigwedge_{d \in \mathcal{D}} \bigwedge_{1 \leq \cdot \leq T} x_d^{\downarrow,\cdot} < W \wedge y_d^{\downarrow,\cdot} < H \wedge x_d^{\uparrow,\cdot} < W \wedge y_d^{\uparrow,\cdot} < H.$$

The lower bound of 0 is inherently enforced by the data type of the variables.

Another constraint that needs to be satisfied in order to have a "well-formed" droplet is to make sure that the lower left corner actually is lower and on the left-hand side of the upper right corner. This is ensured by the constraints

$$\bigwedge_{d \in \mathcal{D}} \bigwedge_{1 \leq \cdot \leq T} x_d^{\downarrow,\cdot} \leq x_d^{\uparrow,\cdot} \wedge y_d^{\downarrow,\cdot} \leq y_d^{\uparrow,\cdot}.$$

Enforcing the Source and Target Locations

Source and target locations must be occupied by the droplets at the beginning and at the end of the experiment, respectively. The droplets are put on their source locations by explicitly setting

$$\bigwedge_{d\in\mathcal{D}} p_d^1 = p_d^*.$$

To ensure that the droplets reach their target locations, the constraints

$$\bigwedge_{d\in\mathcal{D}} \left(\bigvee_{t=1}^{T} p_d^t = p_d^{\dagger}\right)$$

are added to the SMT instance.

These constraints are entirely analogous to the ones from the routing constraints from (3.4) and (3.5).

Modeling the Droplet Movement

Droplets must not arbitrarily occupy a cell position. The same model assumption as for conventional DMFBs applies. That means that a droplet's corner may be present at a position p, only if it was present in the neighborhood of this position in the previous time step. This is captured in the constraints

$$\bigwedge_{d\in\mathcal{D}}\bigwedge_{t=2}^{T} \left|x_d^{\downarrow,} - x_d^{\downarrow,t-1}\right| \le 1 \wedge \left|y_d^{\downarrow,} - y_d^{\downarrow,t-1}\right| \le 1 \qquad (5.4)$$

and

$$\bigwedge_{d\in\mathcal{D}}\bigwedge_{t=2}^{T} \left|x_d^{\uparrow,} - x_d^{\uparrow,t-1}\right| \le 1 \wedge \left|y_d^{\uparrow,} - y_d^{\uparrow,t-1}\right| \le 1 \qquad (5.5)$$

for the movements of the lower left corner and upper right corner, respectively. The constraints above allow for every corner of the droplet to move one cell in any direction, including diagonal movement.

Different Droplet Velocities

The velocity of the moving droplets may vary depending on the droplet's volume. This behavior is modeled as follows. The fastest droplet defines the base clock time and the movement of all other droplets is relative to that droplet. Each droplet is assigned a movement modifier m_d that takes this relative velocity differences into account. The modifier for the fastest droplets is 1. Using this, the slower movement of droplets is enforced by assuming that a droplet stays on the same location for m_d time steps before reaching a new position. The corresponding constraint for the x coordinate of the lower left corner of a droplet d is given by

$$\bigwedge_{i=1}^{m_d-1} x_d^{\downarrow,t-i} = x_d^{\downarrow,t-m_d}$$

for $t = 1 + k \cdot m_d$ and $k = 1, 2, \ldots$ such that $t \leq T$. This divides the time steps into time frames of size m_d and a remainder that has less than m_d time steps. The constraints ensure that within a time frame, defined by the movement modifier, a droplet must stay at a given position. This means that only every $(k \cdot m_d)$th time step, the droplet d can change its location.

Note that in the case of $m_d = 1$ this constraint has no effect for that particular droplet.

For the sake of brevity, the constraint of a single coordinate only is shown; the other constraints are analogous. Also, necessary constraints that prevent droplets to move more than one position in a single time step are not shown as they are of technical nature only. This can occur in the last time steps when T is not of the form $1 + k \cdot m_d$ for a droplet d.

Example 5.7 Consider a droplet d with a movement modifier $m_d = 3$. The constraint for the x-coordinate of the lower left corner for time step $t = 1 + 1 \cdot m_d = 4$ is then given by

$$x_d^{\downarrow,3} = x_d^{\downarrow,1} \wedge x_d^{\downarrow,2} = x_d^{\downarrow,1}.$$

This ensures that the droplet remains on its position at time steps 1, 2, and 3.

Aspect Ratio Constraint

Besides velocity, also the droplet aspect ratios may change. As discussed, due to the constant volume of liquid, droplets cannot arbitrarily change their aspect ratios. To incorporate this into the SMT formulation, for every droplet d an *aspect ratio library* R_d is defined. It contains all the aspect ratios, the droplet may assume.

Example 5.8 To allow the droplet d to change its ratio from $3 : 3$ to $4 : 4$, the droplet aspect ratio library $R_d = \{(3, 3), (3, 4), (4, 4)\}$ could be used. Note that this library forces the droplet to expand vertically first as the ratio $4 : 3$ is not part of the library.

The allowed width and height of a given ratio $r \in R_d$ are denoted by r_w and r_h, respectively.

To enforce a droplet to assume the allowed ratios only, the specified widths and heights are added to the lower left corner of the droplet to make sure that the resulting coordinate is equal to the droplet's upper right corner. The corresponding constraint is given by

$$\bigwedge_{d \in \mathcal{D}} \bigwedge_{1 \leq \leq T} \left(\bigvee_{r \in R_d} x_d^{\downarrow \cdot} + r_w = x_d^{\uparrow \cdot} \quad \wedge \quad y_d^{\downarrow \cdot} + r_h = y_d^{\uparrow \cdot} \right). \tag{5.6}$$

If one does not want to enforce a dedicated droplet aspect ratio library but to allow arbitrary ratios for the droplets, including extreme cases such as $1 : n$ for a large n or $1 : 1$, as done in [Che+11], the constraints (5.6) can simply be omitted.

Note that the actual change in the droplet aspect ratio results from the movement of at least one of the droplet's corners. The constraints (5.6) merely ensure that the resulting droplet is always of valid aspect ratio.

Respect Blockages

Droplets should avoid moving into blockages. In contrast to routing on conventional DMFBs, droplets should further keep a distance from the blockages, effectively having a safety region around the blockages. To ensure that no droplet moves too close to a blockage, the following constraints check that no corner of the droplet is within the rectangle defined by safety region surrounding a blockage.

$$\bigwedge_{d \in \mathcal{D}} \bigwedge_{b \in \mathcal{B}} \bigwedge_{1 \leq t \leq T} \left(x_b^{\uparrow} + dist < x_d^{\downarrow \cdot} \vee y_b^{\uparrow} + dist < y_d^{\downarrow \cdot} \right.$$
$$\left. \vee x_b^{\downarrow} - dist > x_d^{\uparrow \cdot} \vee y_b^{\downarrow} - dist > y_d^{\uparrow \cdot} \right) \tag{5.7}$$

Here, $dist$ defines the safety margin (usually 1) and $\mathcal{B}$ is the set of all blockages. This safety region can be thought of as static fluidic constraints for blockages.

Example 5.9 Consider again the running example from Example 5.5. The blockage is at location $((4, 1), (5, 2))$. Fixing the safety margin to 1, the corresponding constraint for the first time step is

$$5 + 1 < x_1^{\downarrow,1} \vee 2 + 1 < y_1^{\downarrow,1} \vee 4 - 1 > x_1^{\uparrow,1} \vee 1 - 1 > y_1^{\uparrow,1}$$
$$5 + 1 < x_2^{\downarrow,1} \vee 2 + 1 < y_2^{\downarrow,1} \vee 4 - 1 > x_2^{\uparrow,1} \vee 1 - 1 > y_2^{\uparrow,1}.$$

Fluidic Constraints

Again, the droplets must not violate the fluidic constraints. As with conventional DMFBs the constraints do not apply to droplets belonging to the same net. In contrast to the conventional biochips, it may be necessary to increase the distance droplets have to keep from each other beyond a single cell. For doing so, the idea of a parameterized safety region around droplets is used to formulate the fluidic constraints.

The corresponding constraint is almost identical to the blockage constraint in (5.7). In the static fluidic constraint case, the current droplet d serves as the "blockage." The constraints are given by

$$\bigwedge_{n \in N} \bigwedge_{\substack{d \in n \\ d' \in \mathcal{D} \setminus n}} \bigwedge_{1 \le t \le T} \left(\begin{array}{c} x_d^{\uparrow,} + dist < x_{d'}^{\downarrow,t} \vee y_d^{\uparrow,} + dist < y_{d'}^{\downarrow,t} \vee \\ x_d^{\downarrow,} - dist > x_{d'}^{\uparrow,t} \vee y_d^{\downarrow,} - dist > y_{d'}^{\uparrow,t} \end{array} \right)$$

The constraints for the dynamic fluidic are almost identical. The only difference is that instead of t, $t - 1$ is used for the droplet that serves as the blockage.

5.5.4.2 Extensions

The problem described in this section is the same as the one from Chap. 3 in the sense that only the plain movements of droplets are considered. In principle, some of the extensions to this routing presented earlier in this chapter can also be used for MEDA biochips. These extensions can be summarized as follows:

Adding temporal aspects to the formulation is quite simple. Temporal blockages are trivially realized by adjusting the time steps used in (5.7) for each blockage. For droplets to be able to spawn, a virtual cell has to be introduced. This cell serves as a valid position for a droplet to be on, indicating the droplet's absence on the chip. The issue with this extension is that it introduces a lot of checks for special cases in the constraints. Conceptionally though, this is a very simple change.

Aging-aware routing can be realized by adding constraints similar to (5.2) and (5.3). The issue here is that one has to tediously add helper variables for each cell on the biochip that counts the amount of usage. As no data on the longevity of MEDA biochips has been reported so far, this extension remains rather theoretical.

Whether different cell forms actually would have an impact on the formulation of the proposed routing solution is an open question. Most likely, droplets would still be in rectangular shapes. This would mean that only the actuation sequences for the concrete physical realization of the biochip need to be adjusted. The routing itself could still be solved using the proposed method.

As MEDA biochips can, due to the technical realization, inherently directly address every single electrode individually, the problem of pin assignment does not occur. Therefore, all pin-related extensions are not applicable for MEDA biochips. If in the future pin assignment becomes an issue, it can be added to the formulation with the same remark on the tediousness as with the aging-aware routing.

5.5.5 *Experimental Results*

The proposed routing methodology allows for an evaluation of the benefits of MEDA biochips (compared to conventional biochips) as well as an *exact* comparison to previously proposed solutions.

To evaluate the benefits of the new technology, the previously used routing benchmarks are taken and scaled up to reflect the high density of micro-electrode cells. Routing results with different configurations are obtained and the influence of the configurations on the quality of the routes is discussed.

The minimal routing results of the proposed methodology are then compared against solutions from previous work.

All experiments have been run on a machine with four Intel Xeon CPUs at 3.50 GHz and 32 GB RAM running Fedora 22.

5.5.5.1 Used Benchmarks

The four benchmark sets used before are considered again. In order to create more realistic benchmarks, they are scaled by the factor 3 in both x and y-direction. The scaling approach has been taken from [Che+11].

An example for a scaled benchmark is shown in Fig. 5.14 where sub-problem 2 of the in-vitro1 benchmark set is displayed. This figure illustrates the main problem with using the benchmark set in a MEDA context: using a safety region around blockages can lead to unsolvable routing problems. The droplet in the right-most

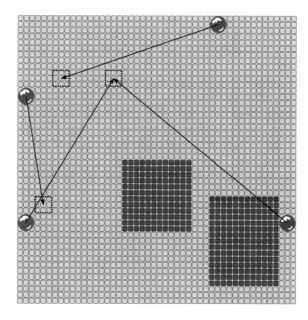

Fig. 5.14 Scaled-up version of sub-problem 2 of the in-vitro1 benchmark set

column has a starting position that immediately violates the blockage constraints. Therefore, in all of the following experiments, the value of *dist* for the blockages is set to zero.

Furthermore, the benchmarks are unrealistic in the sense that they were generated from real protocols being executed on conventional DMFBs. The same protocols, executed on a MEDA biochip, might have created completely different routing problems.

While the (scaled) routing problem sets do serve as a good basis for benchmarking algorithms on MEDA biochips, as is, for example, done in [Che+11] and [Li+16], there is a need for dedicated MEDA benchmarks.

5.5.5.2   Comparison with Conventional DMFBs

As discussed above, MEDA biochips allow for a more advanced droplet manipulation. This results in MEDA biochips being capable of changing the droplets' shapes, and moving droplets diagonally. Furthermore, droplets of different volumes may move at different velocities.

Using the proposed methodology, it is possible to evaluate the precise effects of these advanced droplet manipulation capabilities. The evaluation is done by comparing the minimal number of time steps obtained by the exact routing approach for conventional DMFBs from Chap. 3 to the minimal number of time steps obtained by the proposed approach for MEDA biochips. The routing problems are solved using different configurations. All combinations of (not) using diagonal movements and (not) allowing changes in the droplets' shapes are evaluated. No evaluation using different velocities is performed as they do not add a completely new routing behavior. Different droplet velocities can be simulated on conventional DMFBs by simply keeping droplets on their position for multiple time steps. The main difference in the routing result will be an increase in the number of necessary time steps.

Even though arbitrary aspect ratio changes are not possible in the physical world it is still worthwhile to produce data for this setting. This setting represents the most flexible case when considering changes in the droplet aspect ratio. This setting, therefore, serves as an upper bound on the possible gains using droplet morphing.

Tables 5.4 and 5.5 present the results for routing without diagonal movement and with diagonal movement, respectively. The results are normalized by computing the equivalent arrival times as proposed in [Che+11], hence the floating point values. This allows to compare the proposed method against the results obtained from conventional DMFB routing. This is further necessary as in [Che+11] only normalized data is published.

As can be seen, using a conventional setup, meaning that no diagonal movement is allowed and restricting the droplet aspect ratios leads to results that are almost identical to conventional biochips. But the more "freedom" is allowed by allowing diagonal movements or arbitrary droplet aspect ratios, the better, in the sense of smaller, the number of needed time steps. The capability of diagonal movements

Table 5.4 Comparison to conventional DMFBs with disabled diagonal movement

Benchmark	Classical DMFB		Restricted aspect ratios			Unrestricted aspect ratios		
	max T	avg. T	max T	avg. T	dur (s)	max T	avg. T	dur (s)
in-vitro1	19.00	12.82	18.33	12.00	725	18.33	12.00	1274
in-vitro2	17.00	11.07	16.33	10.40	880	16.33	10.40	3246
protein1	21.00	16.28	20.33	15.62	4697	20.33	15.62	16,558
protein2	21.00	10.53	20.33	9.82	1683	19.00	9.77	6004

Restricted droplet aspect ratios means that the shape library $\{(3, 3), (3, 4), (4, 3), (4, 4)\}$ is used; otherwise, arbitrary droplet aspect ratios can be used

Table 5.5 Comparison to conventional DMFBs with diagonal movement

Benchmark	Classical DMFB		Restricted aspect ratios			Unrestricted aspect ratios		
	max T	avg. T	max T	Avg. T	dur (s)	max T	Avg. T	dur (s)
in-vitro1	19.00	12.82	15.33	9.42	155	15.33	9.42	582
in-vitro2	17.00	11.07	12.00	7.64	88	12.00	7.64	170
protein1	21.00	16.28	18.00	12.85	917	18.00	12.81	2717
protein2	21.00	10.53	17.66	8.06	729	17.00	7.88	1070

Restricted droplet aspect ratios means that the shape library $\{(3, 3), (3, 4), (4, 3), (4, 4)\}$ is used; otherwise, arbitrary droplet aspect ratios can be used

has the most significant impact, while the impact of different droplet aspect ratios is moderate. The latter is not surprising as none of the benchmarks contains "tunnels" of size smaller than the original droplets' sizes. The potentially beneficial case of forming a snake-like droplet to "squeeze" through these tunnels does not occur. As already discussed in Sect. 5.5.5.1, dedicated MEDA routing benchmarks are needed.

Restricting the droplet aspect ratio is a reasonable choice as the droplets' volumes do not change during routing. This means that a model allowing to change the droplet aspect ratio from 2 : 1 to 10 : 23 or from 6 : 7 to 1 : 1 would result in droplets becoming arbitrarily thin or thick, respectively.

5.5.5.3   Comparison to Previous Work

Previous work only considered an approximation of the actual routing time [Li+16] or provided a heuristic solution for the determination of the actual routing [Che+11]. For both approaches it is unknown how far these results are from the actual minimum. The proposed exact routing solution is used to evaluate both approaches with respect to the number of time steps used in their solutions. The corresponding numbers are provided in Table 5.6.

The results show that the work of [Li+16] is overly optimistic. For all benchmarks, a smaller number of time steps are approximated than required by a minimal routing. This under-approximation is caused by the fact that only distances are used to estimate the number of time steps. Blockages or cells to be avoided in order to respect the fluidic constraints are not considered at all. In contrast, the work

Table 5.6 Comparison of the exact MEDA routing to previous work

Benchmark	Approximation [Li+16]		Heuristic [Che+11]		Proposed	
	max T	Avg. T	max T	Avg. T	max T	Avg. T
in-vitro1	12.33	8.2	15.33	9.55	15.33	9.42
in-vitro2	10.67	6.33	11.33	7.29	12.00	7.64
protein1	15.67	9.36	18.67	12.30	18.00	12.81
protein2	12.33	5.25	16.67	7.48	17.00	7.88

of [Che+11] generates much better results. The proposed methodology shows that routings heuristically determined by the A*-based approach are already close to the optimum. Note that we explicitly model the change in aspect ratio. This explains the shorter max T of 11.33 from [Che+11] for the in-vitro2 experiment compared to the value of 12.00 determined by the proposed method.

5.6 Summary

In this section, the routing solution from Chap. 3 and the pin assignment solution from Chap. 4 were combined in order to solve the *pin-aware* routing problem which aims at minimizing both the time steps necessary to route droplets and the number of pins used to realize the routes.

It turns out that the formulation of the pin-aware routing problem is very versatile and easily extendable. This chapter illustrates this by presenting representative scenarios for different use cases.

The pin-aware routing itself can already be used to solve different tasks. One task is to move droplets using a given pin assignment, a task that has not been explicitly addressed before in the literature. It also allows to extend the analysis of given pin assignments from the previous chapter by the capability to check whether a given pin assignment allows for certain droplet movements.

One extension is to take temporal aspects into account. This has been done for the first time. This removes the need to split routing problems into sub-problems, creating a more realistic routing problem. Temporal aspects also extend to blockages, removing the constraint that blockages need to be present the whole time. This is a reasonable assumption as blockages usually represent operations that will finish their execution at some point in time. It was shown that these temporal blockages can already lead to shorter routes.

Another aspect that can easily be handled is the aging problem of DMFBs. Each usage of a cell potentially degrades it eventually leading to a total cell failure. The presented solution allows to restrict the number of cells used during an experiment as well as the number of actuations of individual cells.

The generality of the solution even allows to route droplets on DMFBs using arbitrary cell shapes. The formulation itself does not need to be changed, only the input data needs to be adjusted accordingly. This feature of the proposed solution is used to evaluate three different cell shapes proposed in the literature: triangles, squares, and hexagons. With respect to time steps necessary to route droplets, hexagons perform best. The next best routing solutions are achieved by the commonly used squares. Triangles have the worst performance.

The last section of this chapter dealt with a different type of biochip: *micro-electrode-dot-array* (MEDA) biochips. These chips allow for diagonal droplet movement and the change of the shape of the droplets. The MEDA routing problem, at least in a restricted setting, is NP-complete. Many ideas from "conventional" biochips also apply in this new domain. This allowed to formulate an exact SMT formulation that, for the first time, faithfully models the droplet movement and changes in droplet shape. Experiments show that a previous work that used approximations to estimate the routing duration was overly optimistic while another work, using an heuristic approach, already produces close-to-optimal solutions.

Chapter 6
One-Pass Design

6.1 The Design Gap Problem

The design problem consists of multiple steps (see Chap. 1). Design gaps among these steps restrict the effectiveness of the entire DMFB realization. These gaps are a frequent cause for problems like placement or routing failures, revealing the demand for design convergence. These problems will become even more critical with a rapid escalation in the number of assay operations incorporated into a single large-scale DMFB.

The following example illustrates the gap between these individual steps by showing that placement may fail if the preceding binding and scheduling steps are conducted without taking into account the whole synthesis problem.

Example 6.1 Consider again the sequencing graph from Fig. 1.5a from Chap. 1. The constraint is to use a biochip of size 5×5 and a module library containing the following two mixers:

- $Mixer_1$ using an area of size 2×3 and having a completion time of 17 time steps
- $Mixer_2$ using an area of size 3×3 and having a completion time of 15 time steps

An *as soon as possible* scheduling scheme schedules the two mixing operations in parallel. In order to reduce the experiment run-time, in the binding step, the $Mixer_2$ module is chosen for both operations. The number of necessary cells is 18, which is less than the available 25 cells. Nevertheless, it is not possible to place these modules on the biochip without the mixers overlapping (Fig. 6.1 illustrates the situation). If the binding step had taken into account not just the sizes of the mixers but also the geometry of both, the mixers and the biochip, the placement step would not have failed.

This example illustrates the design gap between the different stages in synthesis for DMFBs.

© Springer International Publishing AG, part of Springer Nature 2019
O. Keszocze et al., *Exact Design of Digital Microfluidic Biochips*,
https://doi.org/10.1007/978-3-319-90936-3_6

Fig. 6.1 Placement failure:
at most one 3×3 mixer is
placeable on a 5×5 biochip
without overlapping another
3×3 mixer. The displayed
configuration has the least
possible overlap (indicated by
the highlighted cell in the
center)

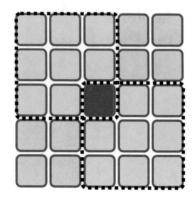

6.2 Proposed Solutions

6.2.1 Heuristic One-Pass Design

This section describes a heuristic approach to the one-pass synthesis problem for
DMFBs (originally proposed in [Wil+15]). The aim is to develop an approach that
is efficient and scalable. A general design methodology for DMFBs is presented
in which all design tasks such as binding, scheduling, placement, and routing are
implicitly conducted in a combined and integrated fashion.

In the following, a heuristic one-pass synthesis scheme is proposed. The idea is
to prevent failures of later design steps in the conventional design flow by trying to
perform all four design steps on nodes of a sequencing graph at once on a node-by-
node basis.

In order to create a scalable and fast procedure, heuristics are used. This does
not guarantee that the determined solutions are minimal with respect to the required
number of time steps. An exact solution for the one-pass synthesis is presented in
Sect. 6.2.2.

The next section describes the general "high-level" flow of the proposed method.
The following sections then explain the individual parts.

6.2.1.1 Main Flow

Given a sequencing graph and a grid of given size, first the design of the DMFB at
time step t is considered. Then, the following steps are performed:

1. A possible operation op to be conducted at time step t is chosen from the given
   sequencing graph.
2. Droplets involved in op are moved towards the position where op is supposed to
   be executed. This requires the consideration of time steps larger than t.

3. An appropriate module realizing *op* is chosen depending on the current status of the biochip.
4. The chosen module is placed on the biochip. This, again, depends on its current status. While doing this, a case distinction is applied:

   (a) If *op* is to be realized by a physical module, existing modules might be used rather than placing a new one.
   (b) If *op* is to be realized by a virtual module, the module is placed for the required number of time steps only. This also may require to consider time steps larger than *t*.

These steps are repeated until either no further operations can be considered (because of dependencies in the sequencing graph which require to wait for the completion of already placed operations in a time step larger than *t*) or heuristics decide that the grid is already too busy and further operations shall not be realized until the completion of currently running operations. Then, *t* is increased by one and the steps from above are executed again. The algorithm terminates when all operations of the sequencing graph have been realized.

Example 6.2 Consider the specification from Fig. 6.2 and a 5×5-grid. First, time step $t = 1$ is considered. Let heuristics decide that the dispensing operations represented by nodes v_1 and v_2 in the sequencing graph from Fig. 6.2a shall be realized first (Step 1). The respective physical modules are chosen (Step 3) and placed (Step 4). In this case, the dispensers are placed at the upper part of the grid. By doing so, the desired droplets are dispensed in time step $t = 1$, meaning that nodes v_1 and v_2 have been realized. The resulting design is depicted in Fig. 6.3a.

In another iteration, it is decided that the mixing-operation represented by the node v_6 shall be realized next (Step 1). For this purpose, the corresponding droplets are moved towards each other and, by this, define the (start) position of the mixing operation (Step 2). In the considered example, this requires $k = 2$ further time steps (see Fig. 6.3a). Then, a virtual module realizing the mixing operation, for example Mixer$_2$, is chosen (Step 3) and placed onto the grid for the next time step (Step 4). This means that starting from time step $t + k + 1 = 4$, cells for this mixing operation are marked as being used for mixing in later time steps. This is illustrated by a dashed rectangle in Fig. 6.3a. These cells can still be used to execute other operations, like droplet movement or storing, before $t = 4$.

According to the module library, Mixer$_2$ needs 15 time steps to perform the mixing operation. Therefore, the corresponding cells of the biochip are assumed to be occupied from time step $t + k + 1 = 4$ until time step $t + k + 15 = 18$.

Following this scheme, all design questions summarized in Chap. 1 are addressed: scheduling (Step 1), routing (through the movements of the droplets in Step 2), binding (Step 3), and placement (Step 4). But in contrast to their separate consideration, all these steps are now conducted in an "interleaved" fashion, meaning that intermediate results can be directly taken into account. This leads to significant benefits as illustrated in the following example.

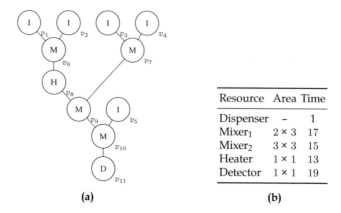

Resource	Area	Time
Dispenser	–	1
Mixer$_1$	2×3	17
Mixer$_2$	3×3	15
Heater	1×1	13
Detector	1×1	19

(a)   (b)

Fig. 6.2 Specification of an experiment used to illustrate the heuristic synthesis approach. (**a**) Sequencing graph. (**b**) Module library

Example 6.3 Following Example 6.2, assume that the algorithm moved forward to time step $t = 4$, in which the mixing operation of node v_6 started its execution.

Furthermore, assume that heuristics already decided to realize the dispensing operations v_3 and v_4 and that the mixing operation v_7 is to be realized next. The current status is shown in Fig. 6.3b. Since the mixing operation v_6 already started its execution, choosing the fastest implementation, namely Mixer$_2$ (see the module library in Fig. 6.2b), would lead to a placement failure as already discussed in Example 6.1. However, following the proposed one-pass-synthesis methodology, the current status on the grid can be considered. Hence, the possible placement failure would be recognized and, instead, a module which fits onto remaining cells is chosen. The only option for doing so is using the Mixer$_1$ module. Following this methodology, a re-spin of previously conducted design tasks can be avoided.

By this, issues like placement failures can directly be addressed or even completely avoided in the first place. Moreover, *all* design decisions from the question which operation shall be considered next (scheduling) to the question where the module is eventually placed (placement) can be made by explicitly considering the current status of the progress of the protocol on the biochip rather than relying on limited information.

The quality of the results of such a scheme as proposed above heavily relies on the applied heuristics. Numerous strategies are possible for the individual steps. This section will present two heuristics. The first one describes how the decision which operation shall be conducted next could be made (this corresponds to a scheme for Step 1). Afterwards, a strategy how to realize a chosen operation is presented (this corresponds to a scheme for Steps 2–4). However, note that many more heuristics are possible and the following descriptions only serve as representatives of how to realize the proposed one-pass-synthesis methodology.

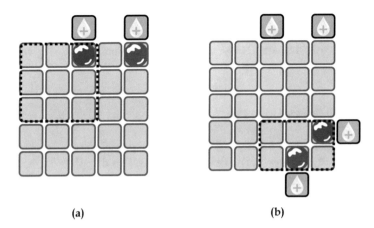

Fig. 6.3 Examples illustrating the main flow of the proposed heuristic one-pass synthesis. (**a**) At time step $t = 1$. (**b**) At time step $t = 4$

6.2.1.2 Choosing an Operation

In order to decide which operation should be considered next, the sequencing graph is analyzed. Only those operations whose predecessors are already completed at the current time step t can be considered. Then, which operation to choose could be decided by means of a prioritization scheme. Possible strategies include decisions based on an analysis of the sequencing graph. This can, for example, be done by

- a breadth-first traversal,
- a depth-first traversal,
- a consideration of the largest number of successors, or
- a purely random scheme.

In case of multiple nodes being eligible for realization, a secondary heuristic must be applied. A choice for such a heuristic can, for example, be to choose a random node or the left-most node first.

Example 6.4 Consider the sequencing graph from Fig. 6.2a. The schemes suggested above would lead to the following list of orderings:

- Breadth-first: $(v_1, v_2, v_3, v_4, v_5, v_6, v_7, v_8, v_9, v_{10}, v_{11})$
- Depth-first: $(v_1, v_2, v_6, v_8, v_3, v_4, v_7, v_9, v_5, v_{10}, v_{11})$
- Successors: $(v_1, v_2, v_6, v_3, v_4, v_8, v_7, v_9, v_5, v_{10}, v_{11})$
- Random: $(v_5, v_3, v_4, v_1, v_7, v_2, v_6, v_8, v_9, v_{10}, v_{11})$

Note that this significantly differs from existing (scheduling) methods since only the order is given by these schemes. The precise time steps in which an operation is actually realized still depends on the current status of the biochip. The prioritization scheme only provides an indication. As an example, an operation for which space would be available according to the current status of the biochip (for example, a detecting operation requiring only one cell) could be realized even if another, larger operation would have a higher priority. Despite the differences, corresponding heuristics for this first step could of course be inspired by existing scheduling methods.

6.2.1.3   Realizing an Operation

In Step 2, droplets are moved towards the position on which the chosen operation op is to be executed. A distinction between operations realized by physical modules and operations realized by virtual modules is made. In the former case, additionally the fact whether a corresponding physical module has already been placed onto the grid needs to be taken into account. If this is the case, design objectives might require that this module has to be re-used again. Then, the position towards which the droplets shall be moved is already known.

If the physical module still has to be placed or if an operation to be realized by a virtual module is considered, a suitable position has to be determined first. For this purpose, the positions of all input droplets for operation op are considered. Then, for each cell of the DMFB, the lowest costs of moving all input droplets to this cell are considered. For this purpose, the direct route is assumed. The corresponding cost function simply is the *Manhattan distance*. It will turn out that this heuristic is a very suitable one. If a cell is currently occupied by another operation op', the maximum of the run-time of op' that is left and the Manhattan distance is used when calculating the costs. Then, the starting position of the operation op is chosen from among the cells with the smallest costs.

Example 6.5 Assume that an operation op with two input droplets (currently at positions $(0, 3)$ and $(3, 1)$) is to be realized. Additionally, assume that cells $(1, 0), (1, 1), (2, 0), (2, 1)$ are currently occupied for another $k = 7$ time steps by a different operation op'. Figure 6.4a shows the costs of using it as a starting point for op for each cell. Based on these values, the starting position is chosen from the five cells with costs of 3: $(0, 1), (1, 2), (2, 2), (2, 3)$, and $(3, 3)$.

Note that the cell $(1, 1)$ has costs of 8 due to the fact that operation op' is currently in progress and can be entered only after $k + 1 = 8$ further time steps. If this operation would be absent, the costs would be 3 and the cell would be considered for use as a starting position.

Note that only the starting position of the module is chosen in this step. This step does *not* choose the concrete module from the module library. This step is performed later on.

Having the starting position, the movement of the droplets towards this position can be conducted. Each droplet is moved along one of the four possible directions (or stalled on its current position). To decide which of these directions should be taken, again the Manhattan distance (with respect to the starting position of the operation) is used. The direction with the lowest value is chosen. Special treatment is necessary if either other droplets or (running) modules are occupying an adjacent

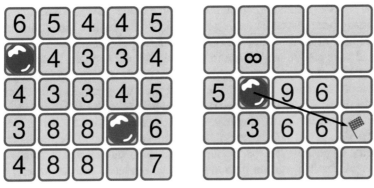

(a) Determining the starting point of a new operation: cost values for each cell

(b) Cost values for starting points of an operation

Fig. 6.4 Realizing an operation: (**a**) determining the starting point and (**b**) moving droplet to a start point

cell. In the first case, the costs of the neighboring cell is set to ∞ avoiding that this direction is taken at all. In the second case, the number of time steps that are necessary to finish the respective operations is added to the Manhattan distance of that cell.

If the direction with the lowest cost value is occupied by another operation, the droplet is stalled until the operation has finished. If multiple cells have the same cost value, the direction is chosen randomly.

Example 6.6 Consider the situation as depicted in Fig. 6.4b. A droplet is to be routed from cell $(1, 2)$ to cell $(4, 1)$. Furthermore, assume that cell $(1, 3)$ is currently occupied by another droplet and, for the next 6 time steps, cells $(2, 1), (2, 2), (3, 1), (3, 2)$ are occupied by another operation. Then, the values ∞, 9, 3, and 5 are determined for the north-, east-, south-, and west-direction, respectively. Note that the costs for the east-direction are calculated by the Manhattan distance plus the time steps in which the currently considered droplet waits for the completion of the operation occupying cell $(3, 3)$.

In this example, the droplet is going to be moved to cell $(1, 1)$, as the cost value of 3 is the lowest.

6.2.1.4 Choosing and Placing a Module

Choosing the concrete module that is used for executing the operations from the module library (Step 3) as well as finally placing it on the DMFB (Step 4) still has to be done.

When choosing a module in Step 3, there is a differentiation between physical modules and virtual modules that is important: for physical modules the decision process needs to take into account that a given module, such as a detecting device, has already been placed and is a good candidate for re-use. Virtual modules are never re-used as the modules "vanish" after finishing its operation.

Finally, the chosen module has to be physically placed on the DMFB in Step 4. Step 2 only determined the starting point for the droplets of the particular operation. This means that this is the cell on which all necessary droplets will be when the operation is to be started. The module determined in Step 3 still needs to be placed on the chip in a non-interfering manner. For detectors of size 1×1 this process is trivial as the starting point covers the whole module. For modules having a bigger geometry, all positions on the "border" are potential starting points, increasing the number of possible module placements. Some modules, such as mixers, can be rotated, transforming a 2×3 module into a 3×2 module, further increasing the search space.

Both steps are dealt with in a similar fashion as has been done in Step 2. Cells and modules are evaluated and given a cost value. These are then used again to determine which cells/module is to be used.

6.2.2 Exact One-Pass Design

In the previous section, a heuristic approach for one-pass synthesis has been presented. Without ground truth values for the exact minimal number of time steps necessary to conduct the experiment, the quality of the heuristic results remains unknown. Hence, an exact one-pass approach is presented in this section (originally proposed in [Kes+14]).

The exact method follows the procedure for exact methods established in the previous chapters. In the first step, the synthesis problem is modeled as a decision problem. Then a cascade of these problems is solved using a SAT solver until the solution using the minimal number of time steps (or the least number of cells) is determined.

Without loss of generality, in order to keep the explanation of this section easy to follow, only mixing and detecting operations are used in the protocols. Furthermore, exactly one detecting device is available for each type of liquid that is to be analyzed.

This section describes the proposed SAT encoding. For a given sequencing graph, a corresponding SAT instance is formulated which is satisfiable if there exists a valid placement and routing realizing the functionality specified by the sequencing graph onto a grid of size $W \times H$ within T time steps.

The DMFB synthesis problem is modeled as follows. A droplet d always is of a certain *liquid type* l. These types are, for example, blood or urine. The set of all liquids used in the protocol is denoted by $\mathcal{L}$. Droplets may only appear on the grid if they are generated either by a dispenser with the correct liquid type or by a corresponding mixing operation. Vice versa, a droplet can only disappear from the grid if it is mixed with another droplet or removed using a sink. Both the number of available dispensers and the number of available sinks may be limited by the

Fig. 6.5 Sequencing graph
used when illustrating the
model for the exact synthesis
approach

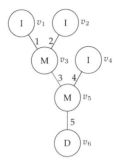

designer (the limits are denoted by $n_{dispensers,l}$ and n_{sinks}, respectively). Modules can arbitrarily be placed onto the grid as long as the cells do not overlap.

All operations op have an a-priori known operation duration dur(op).

All droplets that are designated to be analyzed during the protocol are collected in the set $\mathcal{D}_{det}$; droplets resulting from a mixing operation are collected in the set $\mathcal{D}_{mix}$.

Example 6.7 Consider the sequencing graph displayed in Fig. 6.5. Each of the five edges defines a droplet d, as indicated by the annotations along the edges. The set $\mathcal{D}$ is given by $\{1, 2, 3, 4, 5\}$. The sets $\mathcal{D}_{mix}$ and $\mathcal{D}_{det}$ are $\{3, 5\}$ and $\{3\}$, respectively. Assume that the operations v_2 and v_4 dispense water in order to dilute the liquid dispensed by operation v_1. Then the set $\mathcal{L}$ is given by $\{1, 2\}$.

The general idea of the proposed approach follows the scheme already established in this book. If the corresponding decision problem is unsatisfiable, the grid size and/or the completion time is incremented and the process is repeated. This iteration is stopped when one of the decision problems is satisfiable. Then, a valid placement and routing can be derived from the corresponding solution of the SAT instance. By initially setting the area and/or the completion time to 1 (or any available lower bound) and by iteratively increasing them, minimality is ensured. This scheme has been used throughout this book. Following this scheme, a minimal solution can be determined much faster than by solely enumerating all possible combinations. In Sect. 6.3, a short discussion on using a different scheme for determining the minimal number of time steps is conducted.

6.2.2.1 SAT Variables

To model the DMFB synthesis problem as a SAT problem, Boolean variables are used. These variables are defined as follows:

Definition 6.1 Consider a $W \times H$-grid which is supposed to conduct an experiment in T time steps. The set of all positions $\mathcal{P}$ and the set of droplets $\mathcal{D}$ is defined as in the previous chapters. In this particular setting, $\mathcal{P}$ equals $\{0, \ldots, W - 1\} \times \{0, \ldots, H - 1\}$.

- *Droplets:* The variables $c_{p,d}$ for $p \in \mathcal{P}$, $d \in \mathcal{D}$ and $1 \leq t \leq T$ indicate whether the droplet d is present on cell p in time step t.
- *Mixers* For mixers, no dedicated variables are used. A mixer $m \in \mathcal{M}$ uses the variables $c_{p,m}^t$ for $p \in \mathcal{P}$ and $1 \leq t \leq T$. This automatically ensures that no droplets move through mixers, removing the need to explicitly model blockages.
- *Sinks and Dispensers* Sinks and dispensers are placed next to cells at the edge of the biochip. The variables $sink_p$ and $dispenser_{p,l}$ indicate the presence of a sink or a dispenser (dispensing a droplet of a given liquid type l) next to position $p \in \mathcal{P}$.
- *Detectors* The variables $detector_{p,l}$ with $p \in \mathcal{P}$ represent whether a detector for droplets of liquid type l is placed at position p.

  The variables $detecting_d^t$ with $1 \leq t \leq T$ represent whether the droplet d is processed by a detecting device in time step t.

Example 6.8 Consider the configuration of a 4×4 DMFB shown in Fig. 6.6a. The figure presents a possible placement of droplets, operations, and entities for a time step t. The two droplets 4 and 9 are of liquid types 1 and 2, respectively. At positions $(0, 3)$ and $(3, 1)$ detecting devices capable of analyzing liquids of type 1 and 2, respectively, are placed. The device at position $(3, 1)$ is currently analyzing droplet 9 and, therefore, not visible.

In the lower left corner, two droplets are currently being mixed. The resulting droplet will be droplet 3.

This configuration is represented by the variables listed in Fig. 6.6b.

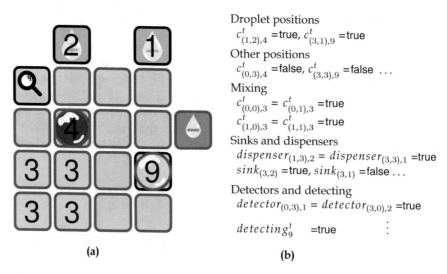

(a)

Droplet positions
$c_{(1,2),4}^t$ =true, $c_{(3,1),9}^t$ =true
Other positions
$c_{(0,3),4}^t$ =false, $c_{(3,3),9}^t$ =false ...
Mixing
$c_{(0,0),3}^t = c_{(0,1),3}^t$ =true
$c_{(1,0),3}^t = c_{(1,1),3}^t$ =true
Sinks and dispensers
$dispenser_{(1,3),2} = dispenser_{(3,3),1}$ =true
$sink_{(3,2)}$ =true, $sink_{(3,1)}$ =false ...

Detectors and detecting
$detector_{(0,3),1} = detector_{(3,0),2}$ =true

$detecting_9^t$ =true   ⋮

(b)

Fig. 6.6 Configuration and encoding of a DMFB configuration at time step t. (**a**) Configuration of the DMFB. (**b**) Corresponding Boolean SAT variables (excerpt)

Given these variables, it is up to the solving engine to assign values for each time step t. To ensure that

- only assignments representing valid placements/routings are chosen and
- the assignment actually realizes the protocol as specified in the sequencing graph,

the variables are further constraints. These constraints are explained in detail in the following.

6.2.2.2 Encoding Consistency, Placement, and Movement

The first constraints that are added ensure that all entities (droplets, physical and virtual modules) are present the correct number of times.

Each position may be occupied by at most one droplet at a time and there also must be at most one physical module at any given position. This is enforced by the constraints

$$\bigwedge_{1 \leq \leq T} \bigwedge_{p \in \mathcal{P}} \left(\sum_{d \in \mathcal{D} \cup \mathcal{M}} c_{p,d} \leq 1 \right),$$

$$\bigwedge_{p \in \mathcal{P}} \left(sink_p + \sum_{l \in \mathcal{L}} dispenser_{p,l} \leq 1 \right), \quad \text{and}$$

$$\bigwedge_{p \in \mathcal{P}} \sum_{l \in \mathcal{L}} detector_{p,l} \leq 1.$$

These consistency constraints have already been used in the previous chapters.

Furthermore, a single droplet may only be present once on the biochip. This is encoded as

$$\bigwedge_{1 \leq \leq T} \bigwedge_{d \in \mathcal{D}} \left(\sum_{p \in \mathcal{P}} c_{p,d} \leq 1 \right).$$

All non-reconfigurable parts such as sinks, dispensers, and detectors also have to be placed the correct number of times. For detectors, it is ensured that for all positions $p \in \mathcal{P}$ and for every liquid type $l \in \mathcal{L}$ of fluids exactly one detector is placed using the constraint

$$\bigwedge_{l \in \mathcal{L}} \left(\sum_{p \in \mathcal{P}} detector_{p,l} = 1 \right).$$

For dispensers and sinks, the constraints are completely analogous. For each position p of the grid and each type of fluid l, we ensure that the desired number of entities ($n_{dispensers,l}$ and n_{sinks}) is placed via the constraint

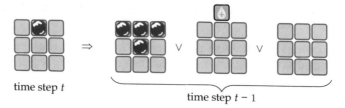

time step t

time step $t-1$

Fig. 6.7 Illustration of the movement constraint

$$\bigwedge_{l \in \mathcal{L}} \left(\sum_{p \in \mathcal{P}} dispenser_{p,l} = n_{dispensers,l} \right) \wedge \left(\sum_{p \in \mathcal{P}} sink_p = n_{sinks} \right).$$

The occurrences of the droplets on the grid (and, therefore, their movement) is encoded as follows. If a droplet d is present at a certain position p at a time step t, one of the following three prerequisites must have been fulfilled:

(a) in the previous time step $t-1$ the droplet was already present at the same cell or one of the neighboring cells (this is identical to the constraints from Chap. 3),
(b) the droplet is next to a dispenser of correct liquid type l which is creating this droplet in the current time step or
(c) the droplet is the result of a mixing operation m. In this case, there must have been a mixing operation on the position p that ended in the previous time step and whose result is the droplet d.

These three conditions are captured in the constraint

$$c_{p,d} \implies \underbrace{\left(\bigvee_{p' \in N(p)} c_{p',d}^{t-1} \right)}_{a)} \vee \underbrace{\left(\bigvee_{p \in \mathcal{P}} dispenser_{p,l} \right)}_{b)} \vee \underbrace{c_{p,m}^{t-1}}_{c)}.$$

Note that for every droplet, in every time step t exactly one of the three parts is **true**. Furthermore, only the parts b) or c) are added to the constraints that are actually applicable. A droplet that is the result of a dispensing operation may not be at a certain position due to the result of a mixing operation.

An illustration of this constraint is provided in Fig. 6.7.

Note that this constraint implicitly encodes dispensing operations, so no constraints for them are necessary.

6.2.2.3 Encoding Operations

In a second series of constraints, the correct execution of the actual operations, such as mixing, detecting, and removal of droplets, is encoded.

Fig. 6.8 The possible
placements for a 2×3
mixing operation on position
$p = (0, 3)$

A mixing operation m requires two droplets, d' and d'', which are, after a given duration of dur(m) time steps, transformed into a new droplet d^*. For this purpose, a fixed number of cells (either a $W \times H$ sub-grid or a $W \times W$ sub-grid) are occupied during this time.

Depending on the cell where the new droplet appears (that is, the "output" cell of the mixing operation) and the module library, several options on the precise placement of these sub-grids exist. The set of such options for a given output cell p is denoted by $M(p, m)$.

Example 6.9 Consider a module library consisting of a 2×3 mixer with a duration of 7 time steps and a 2×2 mixer using 10 time steps. Let $p = (0, 3)$ be the output cell of the mixing operation. As illustrated in Fig. 6.8, there are two sub-grids $\mathcal{P}_1$ and $\mathcal{P}_2$ that could be occupied by the first mixer. These sets are $\mathcal{P}_1 = \{(0, 1), (0, 2), (0, 3), (1, 1), (1, 2), (1, 3)\}$ and $\mathcal{P}_2 = \{(0, 2), (0, 3), (1, 2), (1, 3), (2, 2), (2, 3)\}$. Furthermore, the 2×2 mixer, not shown in Fig. 6.8, can only occupy the sub-grid $\mathcal{P}_3 = \{(0, 2), (0, 3), (1, 2)(1, 3)\}$. These sets of possible placements, together with the corresponding operation durations, yield the set $M(p, m) = \{(\mathcal{P}_1, 7), (\mathcal{P}_2, 7), (\mathcal{P}_3, 10)\}$.

The choice of the sub-grid is only implicitly encoded. To encode the situation that a droplet d^* is the result of the mixing operation $m \in M$ and appears on position p at time step t, the following constraints are sufficient. It needs to be ensured that

(a) the two input-droplets d' and d'' were present in the neighborhood $N(p)$ right before the mixing operation started at time step $t - (\text{dur}(m) + 1)$,
(b) the two input-droplets disappeared right after the start of the mixing operation at time step $t - \text{dur}(m)$, and
(c) one of the possible sub-grids from $M(p, m)$ is chosen, meaning that the precise number of cells that form the mixer is explicitly fixed.

Combining the points above gives the following constraint, for which an illustration is provided in Fig. 6.9.

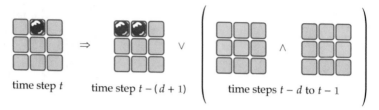

time step t time step $t-(d+1)$ time steps $t-d$ to $t-1$

Fig. 6.9 Illustration of the mixing constraint

$$\bigwedge_{m \in M} \bigwedge_{\text{dur}(m)+2 \leq t \leq T} c_{p,d*}^{t-1} \wedge \neg \bigvee_{p' \in \mathcal{P}} c_{p',d*}^{t-1}$$

$$\implies \bigwedge_{d \in \{d',d''\}} \underbrace{\left(\sum_{p' \in N(p)} c_{p',d}^{t-(\text{dur}(m)+1)} = 1 \wedge \neg \underbrace{\bigvee_{p' \in \mathcal{P}} c_{p',d}^{t-\text{dur}(m)}}_{b)} \right)}_{a)} \wedge$$

$$\underbrace{\bigvee_{(\mathcal{P}_m,\text{dur}(m)) \in M(p,m)} \left(\bigwedge_{t-\text{dur}(m) \leq t' < t} \bigwedge_{p' \in \mathcal{P}_m} c_{p',m}^{t'} \right)}_{c)}$$

Note that some further consistency constraints for the mixing operation are necessary. For example, it must be ensured that no additional stray cells are occupied by the mixing operation and that exactly one of the possible mixer implementations is chosen. However, those constraints are rather technical and very straightforward to implement. As they do not provide further insights into the proposed encoding, they are omitted for clarity.

Next, the constraints ensuring the correct execution of detecting operations are introduced. For a detecting operation to be correctly executed, it has to be ensured that the droplet to be detected and the detector are placed accordingly and that the detection time is considered. More precisely, if a detecting operation is to be executed for droplet d (of type l) starting at the time step t and, at the same time, the droplet d is on position p, then

(a) a corresponding detector det has to be present at position p and
(b) for the duration $\text{dur}(det)$ of the detecting operation, the droplet must not leave the cell.

This is encoded by the constraint

$$detecting_d^t \wedge \neg detecting_d^{t-1}$$

$$\implies \underbrace{detector_{p,l}}_{a)} \wedge \underbrace{\bigwedge_{t \leq t' \leq t+\text{dur}(det)} \left(c_{p,d}^{t'} \wedge detecting_d^{t'} \right)}_{b)}.$$

Finally, the disappearance of droplets needs to be encoded. Droplets may disappear either when they leave the grid through a sink or are used by a mixing operation. Hence, if a cell p was occupied by a droplet d at time step $t-1$, which is not present in the neighborhood $N(p)$ at time step t anymore, then this implies that there is either a) a sink that is reachable from p or b) a mixing operation m began in the neighborhood of p at time step t. The corresponding constraint is given by

$$c_{p,d}^{t-1} \wedge \neg \bigvee_{p' \in N(p)} c_{p',d}^{t} \implies \underbrace{\bigvee_{p' \in N(p)} sink_{p'}}_{a)} \wedge \underbrace{\bigvee_{m \in M} \bigvee_{p' \in N(p)} c_{p',m}^{t} \wedge \neg c_{p',m}^{t-1}}_{b)}.$$

Again, this constraint is adapted depending on whether a droplet is the input of a mixing operation.

6.2.2.4   Encoding the Objective

All constraints outlined above ensure that the solving engine determines variable assignments representing valid placements and routings. However, additionally it has to be ensured that the actually desired operations (as specified by the sequencing graph) are realized. For this purpose, final constraints enforcing their execution need to be added.

The mixing constraints encoded how the droplets are to be mixed but not that any mixing actually takes place. To make sure that every mixing operation is performed, the presence of every droplet is required for at least one time step. This is accomplished by the constraint

$$\bigvee_{1 \leq t \leq T} \bigvee_{d \in \mathcal{D}} \bigvee_{p \in \mathcal{P}} c_{p,d}.$$

In a similar fashion, all detecting operations are triggered. Instead of ensuring that at least one $detecting_d$ variable is true, we require that the sum of all such variables is equal to the duration $dur(det)$ of the corresponding detecting operation det. This is captured in the constraint

$$\bigwedge_{d \in \mathcal{D}_{det}} \left(\sum_{1 \leq t \leq T} detecting_d = dur(det) \right).$$

Combining all constraints introduced in this section, a satisfiability instance is created, which is satisfiable if, and only if, there exists a valid placement and routing realizing the functionality specified by the sequencing graph.

If a satisfiable assignment is returned, the precise placement and routing for the considered problem can be derived from the assignment of the Boolean variables (see Example 6.8). Otherwise, it has been proven that the given sequencing graph cannot be realized with the current restrictions on grid size and completion time.

6.3   Experimental Results

6.3.1   Considered Benchmarks

Two sets of benchmarks are considered. The first set consists of multiplexed in-vitro diagnostic benchmarks taken from [SC08] that are parametrized in the number of samples and reagents used in the protocol. Each of the n samples is to be mixed with each of the m different reagents. The corresponding parametrized sequencing graph is shown in Fig. 6.10. For n samples and m reagents, this generic sequencing graph has a size of $4mn$. The operation completion times for the mixing and detecting operations have been taken from [SC08].

Further benchmarks are taken from the set of benchmarks that are shipped with the tool *MFSimStatic* [Gri+12] which is freely available on the internet. These benchmarks completely specify the geometry of the biochip and the available modules as well as their operation durations.

6.3.2   Implementations

The two approaches discussed in this chapter have been implemented concurrently.

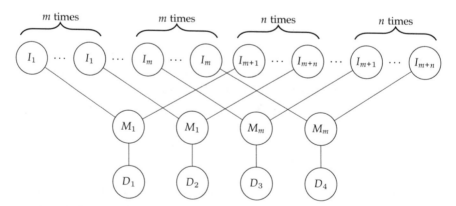

Fig. 6.10 Parametrized sequencing graph for multiplexed in-vitro diagnostics. For n samples and m reagents, the graph has $4mn$ nodes (graphic reproduced and adapted from [SC08, Fig. 12])

For the heuristic approach, a Java implementation was chosen. All results for the heuristic approach have been conducted on a 2.4 GHz Core i7 machine with 8 GB of main memory.

The exact approach has been implemented in Ruby which, given a sequencing graph, a grid size $W \times H$, and a completion time T, generates the satisfiability instance using the SMT-LIB2 format [BST10] extended by a logic for cardinality constraints [Rie+14]. The extension has been implemented on top of the open source toolkit *metaSMT* [Rie+16]. As the solving engine, *Z3* [DB08] was chosen. All experiments for the exact approach have been conducted on a 2.6 GHz Intel Core i5 machine with 8 GB of main memory.

6.3.3 Evaluation of the Solution Length

In a first set of experiments, the heuristic approach has been evaluated using rather large synthesis problems shipped with *MFSimStatic*. Since the proposed heuristic methodology heavily relies on heuristics using random numbers, several runs with different configurations have been performed. In Table 6.1, the best and worst results obtained from 300 different runs are reported in columns min T and max T, respectively. The column dur (s) shows the run-time of the approach for the given benchmark accumulated over all 300 individual runs.

The results clearly show that the applied heuristics and random seeds in fact have a significant impact on the quality of the result. The best and the worst results differ by up to roughly 350%. However, since all runs can be conducted in just a few seconds, it is a valid strategy to simply apply different configurations and, afterwards, take the best result.

Compared to existing approaches for single design tasks, these numbers show an improvement with respect to scalability. For example, the heuristically generated results from [SC08] consider grid sizes of merely up to 2×18—while, at the same time, only addressing the scheduling problem. Significantly larger designs have completely been processed using the approach proposed in this book.

The next set of experiments uses the multiplexed in-vitro diagnostics used in [SC08]. They are used to (a) see how close the results of [SC08] are to the optimum when performing the whole synthesis, not just the scheduling step and (b) evaluate the quality of the heuristic approach with guaranteed optimal results. For this, 19 synthesis problems for different configurations of the grid size and number of samples and reagents are solved.

To achieve comparability to [SC08], the bindings of the mixing operations are fixed according to [SC08, Table I]; the durations are taken from [SC08, Table III]. The results are presented in Table 6.2; #S and #R denote the number of samples and reagents, respectively. As solutions to the synthesis problem are symmetric with respect to the width and height of the grid, the smaller dimension is always displayed first in the table. A solution for a $W \times H$ grid is a valid solution for a $H \times W$ grid by simply "mirroring" all operations on the main diagonal of the grid.

Table 6.1 Experimental results for the DMFB synthesis problem using the heuristic approach

Benchmark	#Nodes	$W \times H$	min T	max T	dur (s)
B1_PCR_Mix	16	2×4	22	73	1.8
B1_PCR_Mix	16	4×4	14	49	1.5
B4_Protein_Mix_Levels_1	28	5×3	35	56	1.6
B4_Protein_Mix_Levels_2	58	5×5	71	138	2.0
B3_2LevelProtein	61	6×6	79	159	2.2
B3_Protein	118	7×7	122	302	3.3
B4_Protein_Mix_Levels_3	118	8×8	121	262	3.3
B4_Protein_Mix_Levels_4	238	10×10	250	405	5.7
	238	12×12	230	400	5.8
	238	14×10	230	367	5.9
	238	14×14	229	448	5.8
B4_Protein_Mix_Levels_5	478	10×10	414	560	9.0
	478	20×20	424	769	11.0
	478	30×30	411	553	13.7
B4_Protein_Mix_Levels_6	958	15×15	792	1002	20.7
	958	20×15	783	973	20.8
	958	20×20	791	995	23.7
	958	25×20	768	1023	24.1
B4_Protein_Mix_Levels_6	958	25×25	791	1065	22.9

As can be seen, the exact approach has a rather long run-time. Therefore, the benchmarks investigated are not identical to the ones from [SC08]. The experiments for two samples and two or three reagents have the same setting except for the grid size. The smaller grid sizes used in this evaluation already show that, in the case of two samples and two reagents, the optimal number of time steps of 15 cannot be achieved on the 2×6 grid used in [SC08]. Instead of the 12 cells of the 2×6 grid, 15 cells on a 3×5 grid are necessary. It is interesting that, even if not on the same grid size, the scheduling result for time steps matches the exact minimal result. For the benchmark for two samples and three reagents, the minimal result uses one time step less.

Compared to the exact results, the heuristic approach hardly uses more than twice the number of time steps in the majority of cases. Considering that the determination of the exact results requires a considerable amount of computational effort, the designs generated by the heuristic methodology are already quite compact—in particular, for an approach that requires very little run-time for their generation.

Table 6.2 Evaluation of the exact and heuristic approach using the multiplexed in-vitro diagnostics benchmarks

			Exact		Heuristic		
#S	#R	$W \times H$	T	dur (s)	T (best)	T (worst)	dur (s)
2	1	4×6	14	51.0	17	29	1.4
2	1	5×5	14	64.1	18	32	1.4
2	1	5×6	14	87.5	18	35	1.5
2	1	6×6	14	185.2	18	31	1.4
2	2	2×5	16	66.9	29	109	1.6
2	2	2×6	16	310.3	28	102	1.5
2	2	3×4	16	173.4	26	94	1.7
2	2	3×5	15	162.8	25	101	1.6
2	2	3×6	15	376.0	25	48	1.5
2	2	4×4	15	277.8	23	48	1.5
2	2	4×5	15	359.3	24	49	1.4
2	2	4×6	15	524.4	24	49	1.5
2	2	5×5	15	503.2	23	51	1.5
2	2	5×6	15	768.2	24	50	1.5
2	2	6×6	15	1262.0	24	87	1.5
2	3	3×6	17	3349.9	33	79	2.1
2	3	4×6	16	1122.9	32	56	1.6
2	3	5×6	16	1874.1	33	65	1.6
2	3	6×6	16	2147.6	33	76	1.7

6.3.4 Evaluating Iteration Schemes

So far, the only iteration scheme for the number of time steps t considered was to start with $t = 1$ or an a-priori known lower bound and then increase t until a solution was determined. Many other iteration schemes are possible. One approach would be to additionally start with an upper bound for which it is known that a solution exists and then perform a binary search between the lower and upper bound.

In order to investigate the feasibility of such an approach, the solving behavior for an instance of the multiplexed in-vitro diagnostics protocol for two samples and three reagents on a 4×6 grid has been analyzed. The minimal completion time for this configuration is 17 time steps. Thus, using the iterative approach, 17 checks are performed in total. In contrast, assuming that the upper bound is known to be 17 then only two checks are necessary; one for 17 time steps and one for 16 time steps to verify that the minimal solution uses 17 time steps.

The solving times for each iteration step are displayed in Table 6.3. Since the durations needed for the first checks of the iterative approach are smaller than the duration for a satisfiable check by two orders of magnitude, the total run-time for both approaches differs only slightly (less than 4%). Similar results have been obtained for other configurations. Therefore, starting an iteration from $t = 1$

Table 6.3 Solving durations for the individual time step SAT instances

T	Iterative		Best case	
	Result	Time	Result	Time (s)
1	UNSAT	0.90 s	–	–
2	UNSAT	2.28 s	–	–
3	UNSAT	3.71 s	–	–
4	UNSAT	5.13 s	–	–
5	UNSAT	6.73 s	–	–
6	UNSAT	8.37 s	–	–
7	UNSAT	10.22 s	–	–
8	UNSAT	12.05 s	–	–
9	UNSAT	13.89 s	–	–
10	UNSAT	16.14 s	–	–
11	UNSAT	18.04 s	–	–
12	UNSAT	20.61 s	–	–
13	UNSAT	23.16 s	–	–
14	UNSAT	26.34 s	–	–
15	UNSAT	29.29 s	–	–
16	UNSAT	63.91 s	UNSAT	63.91 s
17	SAT	3152.45 s	SAT	3152.45 s
Total		3413.22 s		3216.36 s

is the most promising scheme. Even the knowledge of a lower bound does not significantly decrease the overall solving time. Moreover, the upper bound used in an alternative scheme is usually not the optimal case. Therefore, multiple checks for satisfiable instances are performed, possibly even increasing the overall run-time of the approach.

6.3.5 Trade-Off Between Grid Size and Time Steps

As already discussed in Sect. 6.2.2, exact results allow to investigate further aspects of the DMFB synthesis problem. In the following, the trade-off between less time steps or a smaller grid is investigated.

Using the exact approach, the issue has been exemplarily investigated for benchmark configurations of the multiplexed in-vitro diagnostics protocol for different numbers of samples and reagents. Figure 6.11 summarizes selected results. The x-axis denotes the total area (determined by multiplying the width and the height of the grid), while the y-axis represents the completion time.

It can be clearly seen from the results that the larger the grid the less time steps are needed to realize the desired functionality and vice versa. Although this is as expected, these results now allow for a better understanding of how to tackle the multi-objective optimization problem of minimizing both the total area and the

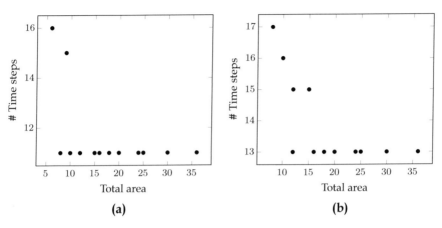

Fig. 6.11 Evaluating the trade-off between grid size and time steps. (**a**) *#samples* = 1, *#reagents* = 2. (**b**) *#samples* = 1, *#reagents* = 3

completion time. The results clearly indicate that increasing the size of the biochip should be favored over allowing more time steps.

As has already been pointed out in Example 6.1 in Sect. 6.1, the geometry of the biochip must be able to fit the modules on itself. This can be seen in Fig. 6.11 where there are "peaks" after there already has been a solution with less cells.

6.4 Summary

This chapter motivates the need for a holistic one-pass design process by giving an example of a design gap problem. This is illustrated by a placement failure due to the decisions of the previous scheduling and binding steps.

To overcome such issues and to achieve exact overall results, two one-pass design methodologies are presented. One solution relies on heuristics trying to schedule, bind, place, and route as many operations and droplets as possible at any time step in an interleaving fashion. This methods aims for short computation time. The other approach is an exact approach that extends the exact formulations of the previous sections to cover all the scheduling, binding, placement, and routing design steps.

Performing exact one-pass design is computationally expensive. This is due to the generally high complexity of the considered problem. Nevertheless, it is possible to determine minimal results for relevant configurations. Though the exact approach may not be used to synthesize large protocols, it still has important applications. First of all, exact results may be used to determine the best solutions for sub-problems of protocols which then are used as building blocks to realize the protocol. For parts that repeatedly occur in a protocol, this may already lead to a significant reduction in execution time.

The more important application is the evaluation of heuristics. Without a proper ground truth, it is impossible to determine the quality of a heuristic. Comparisons between heuristics are not sufficient as they only give relative results.

On the other hand, the heuristic design approach scales well, even for rather large assays. Using enough iterations, results that need only twice as many time steps as the optimal solutions are obtainable. It remains an open question how many iterations should be used or how to decide that a particular solution is good enough to be used.

Chapter 7
Conclusion and Future Work

Many biological assays require a lot of manpower and expensive equipment in the form of highly skilled experts working in large laboratories. The desire to achieve a higher throughput in less time led to the development of different types of robotic devices. These devices also reduce the likeliness of errors due to mistakes caused by users.

To further increase the degree of automation and to reduce the physical size of the device itself as well as the volumes of liquids used during an experiment, different types of microfluidic biochips have been introduced.

The type of biochips considered in this book, *digital microfluidic biochips* (DMFBs), make use of the electrowetting-on-dielectric effect to actuate small volumes of liquids in the form of droplets. These biochips are very versatile and re-usable as well as re-configurable.

The design process for these biochips is usually conducted either manually or using heuristic methods. However, a significant drawback of the heuristic approaches is that they cannot guarantee the optimality of the solutions.

As was shown in the book, the design steps routing and pin assignment, and, therefore, design problems including these steps, are NP-complete. To this end, exact routing and pin assignment solutions have been presented. These solutions, and their combination, do not simply determine optimal solutions for their respective problem. As has been shown, they can be used to solve a large variety of different problems, including the optimization and validation of existing biochips. Furthermore, they also allow for an evaluation of the heuristic approaches from the literature as the exact solutions provide ground truth to compare the heuristic approaches against.

The overall goal when using biochips is to conduct a biological assay. For being able to do so, solutions to all design steps need to be combined to an overall design. Exact solutions to the individual steps in the design process do not necessarily lead to an optimal or even close-to-optimal solution to the whole design problem. It is even possible that these partial solutions cannot be combined at all. To overcome

© Springer International Publishing AG, part of Springer Nature 2019
O. Keszocze et al., *Exact Design of Digital Microfluidic Biochips*,
https://doi.org/10.1007/978-3-319-90936-3_7

this problem of design gaps, this book presented an exact one-pass design solution. This solution is capable of guaranteeing optimality while, at the same time, avoiding all design gap issues.

While the exact one-pass design is a very complex problem, this book not only, for the first time, presented such a holistic approach but also could prove that determining exact solutions is actually feasible.

While the contributions of this book already cover many problems in the design of digital microfluidic biochips, they can still be extended and serve as a starting point for further research. Furthermore, some solutions come with new interesting problems on their own.

The heuristic pin assignment framework presented in this book produced rather mediocre results for the applied heuristics. As low computation time was the main criterion for the choice of heuristics, this is acceptable in the context of this book. However, the presented framework is general enough to be used with many different heuristics and an interesting open problem is to find heuristics that yield better results and to analyze the increase in run-time necessary to achieve these better results.

The routing solution for MEDA biochips already dealt with a new technology. This, again, is only one step in the full design flow. An obvious direction for future work is to extend the routing solution to cover the whole design problem for MEDA biochips.

An open issue for the one-pass design is to incorporate the pin assignment directly into the heuristic and exact design approaches. Furthermore, the exact one-pass design solution does have scalability issues when hundreds of operations need to be performed in an assay. A challenging future work is to create a hybrid solution that determines exact solutions to be used as buildings blocks in a heuristic fashion. This could be a promising solution for larger problems.

The formal model introduced in this book is very general. Some commercial products, such as the NeoPrep System [Illumina], make use of a specialized ring-bus architecture. It would be interesting to see how the findings of this book allow for formulating an exact solution to the design problem on such kind of biochips. An initial work on this has recently been published in [Kes+18].

Appendix A
BioGram: A Dedicated Grammar for DMFB Design

Even though many researchers work in the field of DMFBs, relatively few publications consider the storage format for solutions or input languages as a central theme.

Currently, the state-of-the-art method of describing protocols still is the use of notebooks. In 2010, a first attempt of formalizing experiments has been done in [AT10] where the *BioCoder* language for describing experiments on biochips was proposed. The language's goal is to supply researchers with a means to easily describe their experiments. The language is implemented on top of C++ and cannot be used for visualization. What BioCoder does, though, is to provide the users with a textual representation for the experiment that consists of a combination of strings of user annotations. The user of BioCode has to manually annotate all operations beforehand.

The *Digital Microfluidic Biochip Static Synthesis Simulator* tool [Gri+12] uses a plain text format both for input and output and comes with a visualization for the synthesized experiments. In principle, it is possible to use this tool to visualize results not obtained by the tool itself. Still, the language is very specific and the visualization is not as responsive or interactive as it could be. In order to quickly analyze and debug results of, for example, routing algorithms or different layouts, a more flexible and user-friendly approach is needed to prevent the tools from getting into the way of quickly designing and testing different approaches.

To overcome these issues, the *BioGram* language was designed to describe synthesis and routing solutions for microfluidic biochips that is both, easy to extend and easy to use. In the following, the details of this language are provided (based on [Sto+17]).

BioGram was implemented as an ANTLR [PQ95] grammar (the full EBNF can be found in Fig. A.1) embedded into the proposed visualization tool presented in Appendix B. In general, the design of this language follows the following principles:

Simplicity and Readability: The language is human-readable. This prevents the use of binary files. Furthermore, the language is easy to understand and learn. In fact, the content of a file is completely comprehensible without having to look at the documentation.

© Springer International Publishing AG, part of Springer Nature 2019
O. Keszocze et al., *Exact Design of Digital Microfluidic Biochips*,
https://doi.org/10.1007/978-3-319-90936-3

```
    grammar ::= ⟨grid⟩ | ⟨blockages⟩ | ⟨nets⟩ | ⟨routes⟩ | ⟨meda_nets⟩
               | ⟨meda_routes⟩ | ⟨modules⟩ | ⟨droplets⟩ | ⟨fluids⟩ |
               ⟨pin_related⟩
       grid ::= 'grid' { ⟨position⟩ ⟨position⟩ } 'end'
  blockages ::= 'blockages' { ⟨position⟩ ⟨position⟩ [ ⟨timing⟩ ] } 'end'
       nets ::= 'nets' { ⟨source⟩ { ',' ⟨source⟩ } '->' ⟨target⟩ } 'end'
     source ::= ⟨ID⟩ ⟨position⟩
     target ::= ⟨position⟩
     routes ::= 'routes' { ⟨ID⟩ [ ⟨start_time⟩ ] { ⟨position⟩ } } 'end'
pin_related ::= ⟨pin_assignments⟩ | ⟨pin_actuations⟩ | ⟨cell_actuations⟩
pin_assignments ::= 'pin assignments' { ⟨position⟩ ⟨ID⟩ } 'end'
pin_actuations ::= 'pin actuations' { ⟨ID⟩ ':' { ⟨actuation⟩ } } 'end'
cell_actuations ::= 'cell actuations' { ⟨position⟩ ':' { ⟨actuation⟩ } } 'end'
  actuation ::= '1' | '0' | 'X'
    modules ::= ⟨mixers⟩ | ⟨detectors⟩ | ⟨dispensers⟩ | ⟨sinks⟩
     mixers ::= 'mixers' { ⟨ID⟩ ⟨time_range⟩ ⟨position⟩ ⟨position⟩ } 'end'
 time_range ::= '[' ⟨Int⟩ '-' ⟨Int⟩ ']'
  detectors ::= 'detectors' { ⟨position⟩ [ ⟨spec⟩ ] } 'end'
       spec ::= ⟨Int⟩ [ ⟨ID⟩ ]
      sinks ::= 'sinks' { ⟨ioport⟩ } 'end'
  dispensers ::= 'dispensers' { [ ⟨ID⟩ ] ⟨ioport⟩ } 'end'
     ioport ::= ⟨position⟩ ⟨Direction⟩
    droplets ::= 'droplets' { ⟨ID⟩ ⟨ID⟩ } 'end'
     fluids ::= 'fluids' { ⟨ID⟩ ⟨Description⟩ } 'end'
   position ::= '(' ⟨Int⟩ ',' ⟨Int⟩ ')'
     timing ::= '(' ⟨Int⟩ ',' ( ⟨Int⟩ | '*' ) ')'
```

Fig. A.1 EBNF for the proposed BioGram grammar

Tool agnosticism: A good language is easily usable by many researchers without focusing on specific approaches. In the design process, great care was taken to cover as many use cases as possible without adding anything too specific. The supported use cases contain grids of different layouts, such as non-rectangular grids as used in [ZC12] and [Dat+14] (see Sect. 5.4), temporal blockages (see Sect. 5.2), pin assignments and pin actuations (see Chaps. 4 and 5), as well as the placement of modules as needed in synthesis (see Chap. 6).

Extendability: While the language supports a fixed set of issues, it can easily be extended by simply defining new blocks. This is possible due to the fact that very few aspects of biochips are encoded in nested structures. The design follows the concept of enclosing lists of certain aspects (for example, droplet movement) by an opening keyword and a closing "end" keyword. Furthermore, parts such as timings can easily be re-used when adding new descriptions to the language.

Fig. A.2 Routing solution
provided in BioGram from
Example A.1

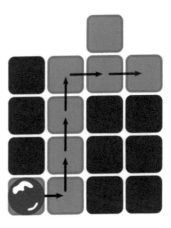

Example A.1 Consider the following routing solution provided in BioGram syntax:

```
grid
    (1,1)  (4,4)
    (3,5)  (3,5)
end
blockages
    (1,2)  (1,4)
    (3,1)  (4,3)
end
routes
    1  (1,1)  (2,1)  (2,2)  (2,3)  (2,4)  (3,4)  (4,4)
end
```

The biochip is defined using two rectangles. The first rectangle defines a 4 × 4
rectangular biochip. The second rectangle adds a single cell on the top row of the
chip.

Two rectangular regions, one on the left side and one on the right side of the chip,
are blocked. A single droplet, with id 1, moves along the path explicitly specified
by listing the cells it travels. The situation is shown in Fig. A.2.

In order to illustrate the extendability of the language, support for MEDA
biochips (see Sect. 5.5) has been added after the initial design of the language was
already finished.

BioGram supports MEDA biochips by simply adding a 'meda' in front of the nets
and routes and slightly changing how positions in the nets and routes are described.
The difference is that a position now is not a tuple describing a single cell but a
rectangle. This is similar to how the grid itself and blockages are described so far.
To the main entry point for the grammar, the production rule ⟨*grammar*⟩, the terms
⟨*medanets*⟩ and ⟨*medaroutes*⟩ are added to incorporate the new aspects of MEDA
biochips. The other additional terms for the grammar are shown in Fig. A.3.

meda_nets ::= 'meda_nets' { ⟨meda_source⟩ { ',' ⟨meda_source⟩ } '->'
⟨meda_target⟩ } 'end'
meda_source ::= ⟨ID⟩ ⟨location⟩
meda_target ::= ⟨location⟩
meda_routes ::= 'meda routes' { ⟨ID⟩ [⟨start_time⟩] ⟨location⟩ } 'end'
location ::= '(' ⟨position⟩ ',' ⟨position⟩ ')'

Fig. A.3 Extension to the initial BioGram grammar as shown in Fig. A.1 to incorporate MEDA biochips

Fig. A.4 Visualization of the MEDA DMFB droplet moving as described in BioGram in Example A.2

Example A.2 Consider the MEDA droplet shape changing and movement provided in BioGram as follows:

```
grid
   (1,1)  (5,5)
end
meda routes
   1 ((1,2),  (3,4))  ((2,2),  (4,3))  ((3,2),  (5,3))
end
```

The example describes a 3 × 3 droplet on a 5 × 5 MEDA DMFB. The three positions on the droplet's route are shown in Fig. A.4. The first thing the droplet does is changing its shape from 3 × 3 to 3 × 2 by moving from the rectangular position $((1, 2), (3, 4))$ to $((2, 2), (4, 3))$. In the next step it simply changes its position without changing the shape again.

The character '#' begins a comment that ends at the end of the line. These comments are for users reading the BioGram file itself. A comment starting with '#!' is treated as an annotation that can be displayed by the BioViz tool introduced in the next section.

Appendix B
BioViz: An Interactive Visualization Tool for DMFB Design

When designing biochips, both structural and behavioral attributes need to be addressed. The structural features consist of the hardware: the cells that are responsible for moving the droplets, the pins that activate these cells, the available dispensers, detectors, and other features all describe the structure of the biochip. The behavior of the chip, on the other hand, is defined by changes of the system's attributes, such as the movement of the droplets across the hardware or the activation and deactivation of the cells.

A visualization that illustrates these attributes should increase the designer's ability to inspect a system's behavior (similar observations on how visualization helps to focus on important structures or properties have already successfully been applied in other domains such as design of conventional software [WLR11], circuits and systems [Sin+00], reversible circuits [Wil+14], or Boolean satisfiability [Sin07]). While the structural features (such as the cell layout) can easily be illustrated using static images, the behavioral attributes require a more dynamic approach. While classic hardware visualization tools have traditionally used static illustration methods even for timed features (such as waveforms for signals), the multi-dimensionality of the behavior of these systems prohibits such an approach. Even the simple case of a single droplet moving on a biochip needs more dimensions to be displayed as its coordinates already require two dimensions to be illustrated. The timing aspect cannot simply be used as one of the coordinates, as is done for waveforms, without resorting to drawing three-dimensionally (as done in [Gri+12]).

In the following a corresponding solution is presented as the visualization tool *BioViz* (originally introduced in [Sto+17]), which is available online at http://www.informatik.uni-bremen.de/agra/bioviz/. The core idea is to provide a dynamic visualization that lets the designer

- easily scroll through time in order to quickly see a particular time step of the protocol or to
- aggregate different states in a single, comprehensive view that allows a quick summary over the total protocol.

O. Keszocze et al., *Exact Design of Digital Microfluidic Biochips*, https://doi.org/10.1007/978-3-319-90936-3

This means that a core requirement of a visualization system for biochips is interactiveness. Designers should be able to quickly browse through different aspects of their design and to easily draw connections between different views. On the other hand, the visualization should be able to provide the basis for further exploitation of the interactiveness.

To show the usability and extendability of BioViz, after introducing the tool itself in the next section, an interesting use case is shown in Sect. B.2. There, BioViz is used to quickly generate routes for droplets using visual feedback.

B.1   The Graphical User Interface

BioViz was built using OpenGL. Instead of providing static illustrations of the system, the core requirement of a responsive, dynamic visualization makes the utilization of such a graphics framework an obvious choice. It allows to implement a dynamic view that can be used to smoothly view changes over time and switch between different views.

As a core framework, the *libgdx* library was used as a wrapper. The remaining logic was implemented using Java, allowing the visualization tool to be used on all major operating systems.

Figure B.1 shows the visualization application with an example system that consists of a 5 × 5 square grid, two dispensers, two droplets and a single sink on the right below one of the dispensers. Using the slider indicated by the "1", one may advance through time in a simple step-through manner. Figure B.2 illustrates this behavior, showing consecutive states of the chip as they would appear to the designer. In addition to the ability to show particular states, an arbitrary number of previous and consecutive states can be aggregated as well using slider indicated by "2". Figure B.3a shows this mode of operation, taking the second time step of the simulation and overlaying the previous and consecutive positions of the droplets by adding arrows to the display.

Figure B.3b illustrates the cells' actuations being displayed using the same paradigm. The visualization can switch between showing distinct states of the design or aggregating several of them to enable designers to quickly see an overview of certain properties. The aggregated information for the number of cell actuations is shown in Fig. B.3c. This is a core idea of the visualization and is available for the various information being displayed for the DMFB.

For the grid, several symbols for different operations are currently embedded into the visualization: cell, blockage, detector, heater, magnet, source, sink, and start and target of nets (see Fig. B.4). More types could easily be added, but as these are the types that are used in current benchmarks, they form a valid base for the visualization to support most use cases.

Another important feature when dealing with designs that change certain properties over time is to enable designers to see how elements behave. When simply displaying discrete states, it may be hard to correctly see how the elements map

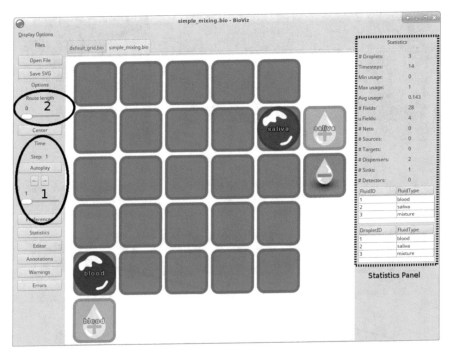

Fig. B.1 Visualization of a simple mixing protocol. The GUI has the controls on the left, the visualization itself in the center, open files at the top and optional statistics on the right-hand side of the window

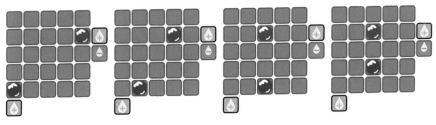

Fig. B.2 The designer can step through the discrete time steps of the design, allowing him to quickly see routing properties and check what happens at which state of the simulation

over time. Therefore, smooth transitions between these states are provided, allowing designers to better understand how the transitions between these states work. Figure B.5 illustrates this transition for droplet movement. However, the same principle applies to, for example, cell colors and viewport shifts. Thus, the designer is constantly being aware of how the states of her design change.

As the designer may freely move and zoom around the chip, he may move the virtual camera far away to get a broad overview over the chip. In order to still provide a usable representation of the system, the visualization switches to a lower level of detail once the parts otherwise become too small to recognize properly [DL05]. Figure B.6 illustrates how at a certain zoom level, the individual elements

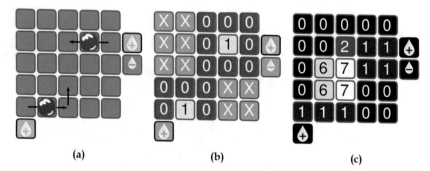

(a)   (b)   (c)

Fig. B.3 (a) Routes can be overlaid in addition to displaying particular states of the design. (b) Distinct information: cell actuations of the second time step. (c) Aggregated information: amount of cell actuation on a color scale from black to white and precise numbers of actuations

Fig. B.4 Currently supported types: default, blockage, detector, heater, magnet, source, sink, start, target

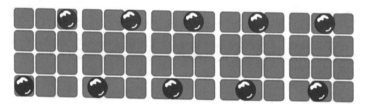

Fig. B.5 States are smoothly transitioned to better illustrate the transitions from one system state to the next

are reduced to simple square boxes that merely hold the color information, still providing the designer with the data that can be drawn from the simulation traces while at the same time avoiding display issues down to the point where individual parts are merely one pixel in size.

BioViz further supports directly editing the loaded file and showing its provided annotations. To simplify debugging, it may show potential warnings (for example, when droplets "jump" over cells) and parsing errors. The statistics panel on the right can be turned off.

As biochips are still subject to fundamental research, designers and researchers strive to report their results in printed form, for example by publishing in research journals. While this, by definition, does not allow any animated properties to illustrate the design properties, it still is an important factor in the development of a given biochip. The implementation supports exporting the current state as a vector graphics file that can be easily embedded into websites or converted to the portable document format (pdf) that can be used in, for example, LATEX-documents.

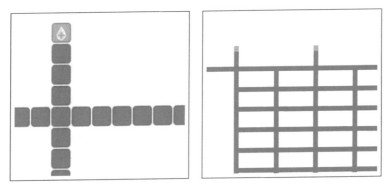

Fig. B.6 Detail is omitted when zooming out to provide a better overview

The visualization was tested with examples from [ZC12, YYC08, CP08] and was used to create the images used in this book. It can handle large layouts and still provide a smooth user experience, allowing designers to interactively inspect their simulation traces and easily see if their algorithms behave as expected.

B.2 Use Case: Interactive Routing

The routing problem and the proposed methodologies (see Chap. 3) are of complex nature. Nevertheless, solving real-world routing problems seems to be easy for humans as they tend to have an intuition for the problem and what routes will be valid. This section presents a human-centered interactive routing tool built on top of BioViz that, starting with an initial guess for routes, allows to easily alter these routes until the design goals are met. This procedure is heavily inspired by Google Maps.

The main idea is to alter routes using mandatory destinations (waypoints) when determining routes for the droplets. Given an initial routing for droplets, that explicitly is allowed to violate design goals, adding waypoints to these routes can be used to resolve the design goal violations. The overall procedure is summarized as follows:

1. Generate an initial (possibly incorrect) routing solution.
2. Check for violations of the design goals. If none are found, return the solution.
3. Identify a droplet whose route is part of a design goal violation.
4. Add a waypoint to its route and recompute its route that moves through the waypoint.
5. Iterate the process until a solution is found.

When a waypoint is added, the route is separated into two new nets: one beginning at the start point of the previous route and ending in the newly added waypoint and one starting in that waypoint and ending in the original end point of the previous route. These nets are then again routed using the simple routing

Fig. B.7 Example of a
BioViz visualization: Two
droplets move along routes
indicated by the
corresponding black arrows.
The violations of the design
rules are clearly visible to the
designer: (a) droplet positions
on which the fluidic
constraints are violated are
highlighted and (b) one
droplet obviously moves
through a blockage

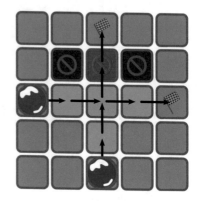

algorithm. When further waypoints are added, the starting and end position for the
new nets may themselves be waypoints that have been added earlier.

B.2.1 Implementation

The tool has been implemented on top of BioViz. BioViz already offers means
to visualize problems in the design and, therefore, is a very natural choice for an
interactive frontend. Figure B.7 illustrates a situation where two common mistakes
and their visualization are shown.

B.2.2 Routing Algorithms

In order to be able to provide the user with initial routes, a simple routing algorithm
is necessary. It is of utmost importance that this routing algorithm comes up with
a solution quickly as this is crucial for the user experience: If a user has to wait
minutes after she added or changed a waypoint, the benefit of an interactive tool
is lost.

Three different routing algorithms of different complexity and run-time
are used:

1. The well-known Dijkstra-Algorithm [Dij59].
2. An extension to the Dijkstra-Algorithm that ensures that no two droplets share
   the same cell in the same time step.
3. An extension to the previous algorithm, enforcing the fluidic constraints [SHC06].

The algorithms are ordered with increasing run-time and quality of the returned
routes. All algorithms ignore blockages as these usually are not that much of an
issue and routes moving droplets through blockages are easy to spot.

Note that there is an even simpler and, therefore, faster algorithm for finding initial routes: Simply create an "L"-shaped route that moves the droplet along one dimension until the coordinate is identical to the target coordinate and then repeats the process in the other dimension. This algorithm is not used as for non-rectangular biochips (see, for example, [ZC12]) almost all generated routes will be invalid, effectively forcing the user to specify every single cell in each route.

Note that the presented algorithms will also work on biochips with alternative cell shapes such as hexagons [SC06b] or triangles [Dat+14] (also see Sect. 5.4).

B.2.3   Case Study

As a case study, the routing problem taken from [HH10], depicted in Fig. B.8a, is used. There are six droplets that need to be routed around a lot of blockages. The droplets are initially routed using the fastest algorithm presented above (Dijkstra ignoring all design goals) using 14 time steps. We can see that droplets 1, 2, 3, and 5 move through blockages. We begin by fixing this issue for droplet 1 and 3 by adding a waypoint at position $(10, 8)$ and $(4, 12)$, respectively. This leads to the situation in Fig. B.8b. While droplets 1 and 3 do not move through blockages any more, droplet 2 and 3 now violate the fluidic constraints. These violations can be partially solved by moving droplets not along the shortest routes but stalling them using detours. This can be seen in Fig. B.8c. Eventually, after adding 10 waypoints in total, a solution not violating any constraints using 16 time steps as shown in Fig. B.8d is found. Note that droplet 1 needs to be re-routed to take some detours in order to find a satisfying solution. In Chap. 3, it has been shown that the shortest possible solution uses 15 time steps. This shows that the human-centered interactive approach is capable of finding good solutions with a reasonable amount of human input.

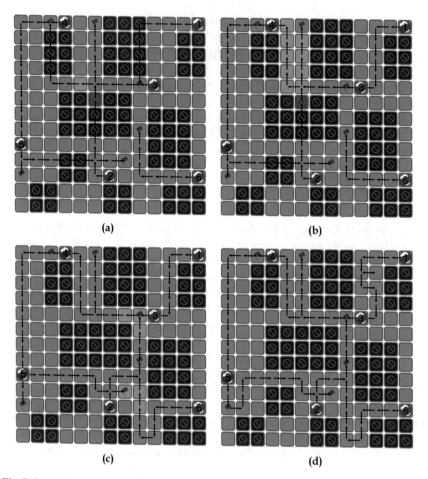

(a)   (b)

(c)   (d)

Fig. B.8 Initial routes to a real-world problem and the solution after adding 10 waypoints. (**a**) Initial solution for the routing of six droplets on a 13×13 biochip using 14 time steps. (**b**) Solution after adding waypoints at $(10, 8)$ and $(4, 12)$ for droplets 1 and 3, respectively. These waypoints add violations of the fluidic constraints. (**c**) Adding waypoints for droplet 2 and 6 enforces to route the droplets on detours to avoid too many violations of the fluidic constraints. This solution now uses 16 time steps. (**d**) Further detours are necessary to find a valid solution using 16 time steps. The simple route of droplet 1 needs to be re-routed considerably

Appendix C
Notation

The following table summarizes the notations used throughout this book. The chapter in which the notation was introduced first is indicated by sub-headings. Symbols that have been introduced in previous chapters are not repeated.

Symbol	Description
	## Chapter 2: Background
$G = (V, E)$	Graph consisting of the set of vertices V and the set of edges E
$d, \mathcal{D}$	A single droplet $d \in \mathcal{D}$ and the set of all droplets $\mathcal{D}$.
$p, \mathcal{P}$	A single position $p \in \mathcal{P}$ and the set of all positions $\mathcal{P} \subset \mathbb{N}^2$.
$p_d^*, p_d^\dagger$	The source and target position of droplet d; $p_d^*, p_d^\dagger \in \mathcal{P}$
$b, \mathcal{B}$	A single blockage $b \in \mathcal{B}$ and the set of all blockages $\mathcal{B} \subset \mathcal{P}$
$n, \mathcal{N}$	A single net $n \in \mathcal{N}$ containing droplet d, and the set of all nets $\mathcal{N}$
n_d	The net to which the droplet d belongs
$\mathcal{D}_n$	The set of all droplets belonging to net n
t, T	A single time step and the maximal time step; $1 \leq t \leq T$
r_d	Route for droplet d; $r_d(t) \in \mathcal{P}$ position at time step t
$N(p)$	Neighboring positions of p that can be reached by a droplet on p in a single time step. It holds that $p \in N(p)$
$N_I(p)$	Neighboring positions of p that must be avoided to ensure the fluidic constraints. The inclusion $N(p) \subset N_I(p)$ must always hold
$\mathbb{A}$,	Set of actuations on (1), off (0), and $don't$ $care$ (X)
$v_p \in \mathbb{A}^T$	Actuation vector of length T for position $p \in \mathcal{P}$.
	## Chapter 3: Routing
I	Edges used for modeling the interference region
$c_{p,d}$	Boolean variable indicating whether droplet d is present on position p in time step t

(continued)

© Springer International Publishing AG, part of Springer Nature 2019
O. Keszocze et al., *Exact Design of Digital Microfluidic Biochips*,
https://doi.org/10.1007/978-3-319-90936-3

(continued)

Symbol	Description
	<div align="center">Chapter 4: Pin Assignment</div>
$A \subset \mathbb{A}^T$	Set of actuation vectors; input to the pin assignment problem
A_i	*Pin set* of actuation vectors belonging to the same pin i
$v_i \circledast v_j$	Binary predicate on actuation vectors indicating the compatibility
$\mathcal{PA}(A, k)$	Pin assignment problem for the actuation vectors A and a maximal number of k pins
$C(G, k)$	Graph coloring problem for the graph G and a maximal number of k colors
pin_v	Integer SMT variable for the actuation vector v
	<div align="center">Chapter 5: Pin-aware Routing and Extensions</div>
act_p^t	Boolean variable indicating whether the electrode at position p at time step t is actuated
$assigned_p$	Pin number that will be assigned to the electrode at position p. This variable is used in the SMT formulation only
u	Upper bound on the overall usage of electrodes.
u_p	Upper bound on the usage of the electrode at position p
W, H	Width and height of a biochip
$a : b$	Aspect ratio of a MEDA droplet
R_d	Droplet ratio library for the droplet d
m_d	Movement modifier for droplet d
$(x^\downarrow, y^\downarrow), (x^\uparrow, y^\uparrow)$	X- and y-coordinate of the lower left (upper right) corner of a MEDA droplet of a blockage on a MEDA biochip
$dist$	Distance around blockages and droplets that must not be entered by other droplets
	<div align="center">Chapter 6: One-Pass Design</div>
v_i	Node of a sequencing graph
op	Some operation that is part of a protocol
$dur(op)$	The duration, in time steps, the operation op takes to finish
$l, \mathcal{L}$	Identifier for a given liquid type $l \in \mathcal{L}$ and set of all liquid types $\mathcal{L}$
$m, \mathcal{M}$	Identifier for a mixing operation $mixer \in \mathcal{M}$ and set of all mixing operations $\mathcal{M}$
$M(p, m)$	All possible bindings for mixing operation m if the input cell is on position p
$n_{sink}, n_{dispenser,l}$	Number of available sinks and dispensers of type l
$\mathcal{D}_{mix}, \mathcal{D}_{det}$	Set of droplets that result from mixing operations and set of droplets that are to be processed by a detecting operation, respectively
$c_{p,m}^t$	Boolean variable indicating whether mixing operation m is being conducted on position p in time step t
$detector_{p,l}$	Boolean variable indicating whether a detecting device capable of analyzing a liquid of type l is present at position p
$sink_p$	Boolean variable indicating whether a sink is placed next to position p
$dispenser_{p,l}$	Boolean variable indicating whether a dispenser of type l is placed next to position p
$detecting_d^t$	Boolean variable indicating whether the droplet d is being analyzed by a detecting device in time step t

References

[ALL] Advanced Liquid Logic. http://www.liquid-logic.com/

[Alb01] Albrecht C (2001) Global routing by new approximation algorithms for multicommodity flow. IEEE Trans Comput Aided Des Integr Circuits Syst 20(5):622–632

[Ali+17] Alistar M et al (2017) When embedded systems meet life sciences: microfluidic biochips for real-time healthcare. Tech. rep. https://www.researchgate.net/profile/Mirela%5C_Alistar2

[AT10] Ananthanarayanan V, Thies W (2010) Biocoder: a programming language for standardizing and automating biology protocols. J Biol Eng 4(1):1–13

[BG06] Bahadur V, Garimella SV (2006) An energy-based model for electrowetting-induced droplet actuation. J Micromech Microeng 16:1494–1503

[BB03] Bailleux O, Boufkhad Y (2003) Efficient CNF encoding of Boolean cardinality constraints. In: Constraint programming 2003. Lecture notes in computer science, vol 2833, pp 108–122

[BYM07] Baird E, Young P, Mohseni K (2007) Electrostatic force calculation for an EWOD-actuated droplet. Microfluid Nanofluid 3(6):635–644

[BST10] Barrett C, Stump A, Tinelli C (2010)The SMT-LIB standard: version 2.0. In: Gupta A, Kroening D (eds) International workshop on satisfiability modulo theories, Edinburgh

[Bha+17a] Bhattacharjee S et al (2017) Dilution and mixing algorithms for flow-based microfluidic biochips. IEEE Trans Comput Aided Des Integr Circuits Syst 36(4):614–627

[Bha+17b] Bhattacharjee S et al (2017) Storage-aware sample preparation using flow-based microfluidic lab-on-chip. In: Design, automation and test in Europe, pp 530–535

[Bie+09] Biere A et al (ed) (2009) Handbook of satisfiability. IOS Press, Amsterdam

[Böh04] Böhringer KF (2004) Towards optimal strategies for moving droplets in digital microfluidic systems. In: International conference on robotics and automation, pp 1468–1474

[CZ05] Chakrabarty K, Zeng J (2005) Design automation for microfluidics-based biochips. ACM J Emerg Technol Comput Syst 1(3):186–223

[Che+11] Chen Z et al (2011) Droplet routing in high-level synthesis of configurable digital microfluidic biochips based on microelectrode dot array architecture. BioChip J 5(4):343–352

[Che+13] Chen Y-H et al (2013) A reliability-oriented placement algorithm for reconfigurable digital microfluidic biochips using 3-D deferred decision making technique. IEEE Trans Comput Aided Des Integr Circuits Syst 32(8):1151–1162

[CYK07] Cheow LF, Yobas L, Kwong D-L (2007) Digital microfluidics: droplet based logic gates. Appl Phys Lett 90(5):054107

[CP08] Cho M, Pan DZ (2008) A high-performance droplet routing algorithm for digital microfluidic biochips. IEEE Trans Comput Aided Des Integr Circuits Syst 27(10):1714–1724

© Springer International Publishing AG, part of Springer Nature 2019
O. Keszocze et al., *Exact Design of Digital Microfluidic Biochips*,
https://doi.org/10.1007/978-3-319-90936-3

[Dat+14] Datta P et al (2014) A technology shift towards triangular electrodes from square electrodes in design of Digital Microfluidic Biochip. In: International conference on electrical and computer engineering, pp 1–4

[De +12] De Leo E et al (2012) Networked Labs-on-a-Chip (NLoC): introducing networking technologies in microfluidic systems. J Nano Commun Netw 3(4):217–228

[De +13] De Leo E et al (2013) Communications and switching in microfluidic systems: pure hydrodynamic control for networking Labs-on-a-Chip. Trans Commun 61(11):4663–4677

[DB08] De Moura L, Bjørner N (2008) Z3: an efficient SMT solver. In: Tools and algorithms for the construction and analysis of systems. Springer, Berlin, pp 337–340. Z3 is available at https://github.com/Z3Prover/z3

[DL05] Deussen O, Lintermann B (2005) Level-of-Detail. In: Digital design of nature. X.media.publishing. Springer, Berlin, pp 181–200

[Dij59] Dijkstra EW (1959) A note on two problems in connexion with graphs. Numer Math1(1):269–271

[DYH15] Dinh TA, Yamashita S, Ho T-Y (2015) An optimal pin-count design with logic optimization for digital microfluidic biochips. IEEE Trans Comput Aided Des Integr Circuits Syst 34(4):629–641

[Don+14] Donvito L et al. (2014) On the assessment of microfluidic switching capabilities in NLoC networks. In: International conference on nanoscale computing and communication, p 19

[Don+15] Donvito L et al (2015) μ-NET: a network for molecular biology applications in microfluidic chips. Trans Netw 24:2525–2538

[ES06] Eén N, Sörensson N (2006) Translating pseudo-boolean constraints into SAT. J Satisfiability Boolean Model Comput 2:1–26

[Fai+07] Fair RB et al (2007) Chemical and biological applications of digital-microfluidic devices. Design Test 24(1):10–24

[FHK03] Fan S-K, Hashi C, Kim C-J (2003) Manipulation of multiple droplets on N×M grid by cross-reference EWOD driving scheme and pressure-contact packaging. In: International conference on micro electro mechanical systems. MEMS, pp 694–697

[FM11] Fidalgo LM, Maerkl SJ (2011) A software-programmable microfluidic device for automated biology. Lab Chip 11(9):1612–1619

[FFW13] Fobel R, Fobel C, Wheeler AR (2013) DropBot: an open-source digital microfluidic control system with precise control of electrostatic driving force and instantaneous drop velocity measurement. Appl Phys Lett:102(19). http://microfluidics.utoronto.ca/dropbot/

[Fré+02] Frénéa M et al (2002) Design of biochip microelectrode arrays for cell arrangement. In: International special topic conference on microtechnologies in medicine biology, pp 140–143

[GJ79] Garey MR, Johnson DS (1979) In: Klee V (ed) Computers and intractability. Books in the mathematical sciences, vol 29. W.H. Freeman and Company, New York

[OD] Gaudi Lab: OpenDrop. http://www.gaudi.ch/OpenDrop/

[Gri+17a] Grimmer A et al (2017) A discrete model for networked labs-on-chips: linking the physical world to design automation. In: Design automation conference, p 50

[Gri+17b] Grimmer A et al (2017) Close-to-optimal placement and routing for continuous-flow microfluidic biochips. In: Asia and South Pacific design automation conference, pp 530–535

[Gri+17c] Grimmer A et al (2017) Verification of networked labs-on-chip architectures. In: Design, automation and test in Europe, pp 1679–1684

[Gri+18a] Grimmer A et al (2018) Design of application-specific architectures for networked Labs-on-Chips. IEEE Trans Comput Aided Des Integr Circuits Syst 37:193–202

[Gri+18b] Grimmer A et al (2018) Sound valve-control for programmable microfluidic devices. In: Asia and South Pacific design automation conference

[GB12] Grissom D, Brisk P (2012) Path scheduling on digital microfluidic biochips. In: Design automation conference, pp 26–35

[Gri+12] Grissom D et al (2012) A digital microfluidic biochip synthesis framework. In: VLSI of System-on-Chip, pp 177–182

[HGW17] Haselmayr W, Grimmer A, Wille R (2017) Stochastic computing using droplet-based microfluidics. In: International conference on computer aided systems theory

[HZC10] Ho T-Y, Zeng J, Chakrabarty K (2010) Digital microfluidic biochips: a vision for functional diversity and more than moore. In: International conference on computer aided design, pp 578–585

[Hu+13] Hu K et al (2013) Fault detection, real-time error recovery, and experimental demonstration for digital microfluidic biochips. In: Design, automation and test in Europe, pp 559–564

[Hu+14] Hu K et al (2014) Control-layer optimization for flow-based mVLSI microfluidic biochips. In: International conference on compilers, architecture and synthesis for embedded systems, pp 1–10

[HH09] Huang T-W, Ho T-Y (2009) A fast routability and performance-driven droplet routing algorithm for digital microfluidic biochips. International conference on computer design, pp 445–450

[HH10] Huang T-W, Ho T-Y (2010) A two-stage ILP-based droplet routing algorithm for pin-constrained digital microfluidic biochips. In: International symposium on physical design, pp 201–208

[HHC11] Huang T-W, Ho T-Y, Chakrabarty K (2011) Reliability-oriented broadcast electrode-addressing for pin-constrained digital microfluidic biochips. In: International conference on computer aided design, pp 448–455

[HLC12] Huang J-D, Liu C-H, T-W Chiang (2012) Reactant minimization during sample preparation on digital microfluidic biochips using skewed mixing trees. In: International conference on computer aided design. ACM, New York, pp 377–383

[HSC06] Hwang W, Su F, Chakrabarty K (2006) Automated design of pin-constrained digital microfluidic arrays for Lab-on-a-Chip applications. In: Design automation conference, pp 925–930

[Illumina] Illumina: NeoPrep Library System. https://www.illumina.com/systems/neoprep-library-system.html

[JBM10] Jensen EC, Bhat BP, Mathies RA (2010) A digital microfluidic platform for the automation of quantitative biomolecular assays. Lab Chip 10(6):685–691

[KWD14] Keszocze O, Wille R, Drechsler R (2014) Exact routing for digital microfluidic biochips with temporary blockages. In: International conference on computer aided design. ICCAD, pp 405–410

[Kes+14] Keszocze O et al (2014) Exact one-pass synthesis of digital microfluidic biochips. In: Design automation conference. DAC, pp 142:1–142:6

[Kes+15] Keszocze O et al (2015) A general and exact routing methodology for digital microfluidic biochips. In: International conference on computer aided design. ICCAD, pp 874–881

[Kes+17] Keszocze O et al (2017) Exact routing for micro-electrode-dot-array digital microfluidic biochips. In: Asia and South Pacific design automation conference. ASP-DAC

[Kes+18] Keszocze O et al (2018) Exact synthesis of biomolecular protocols for multiple sample pathways on digital microfluidic biochips. In: International conference on VLSI design

[Knu15] Knuth DE (2015) Satisfiability. Vol 4, fascicle 6. The art of computer programming. Addison Wesley, Boston

[Lai+15] Lai KY-T et al (2015) A field-programmable lab-on-a-chip with built-in self-test circuit and low-power sensor-fusion solution in 0.35 μm standard CMOS process. In: Asian solid-state circuits conference, pp 1–4

[LYL15] Lai KY-T, Yang Y-T, Lee C-Y (2015) An intelligent digital microfluidic processor for biomedical detection. J Signal Process Syst 78(1):85–93

[Li+16] Li Z et al (2016) High-level synthesis for micro-electrode-dot-array digital microfluidic biochips. In: Design automation conference, p 146

[LC10] Lin CC-Y, Chang Y-W (2010) ILP-based pin-count aware design methodology for microfluidic biochips. IEEE Trans Comput Aided Des Integr Circuits Syst 29(9):1315–1327

[Liu+04] Liu RH et al (2004) Self-contained, fully integrated biochip for sample preparation, polymerase chain reaction amplification, and DNA microarray detection. Anal Chem 76(7):1824–1831

[LC13] Luo Y, Chakrabarty K (2013) Design of pin-constrained general-purpose digital microfluidic biochips. IEEE Trans Comput Aided Des Integr Circuits Syst 32(9):1307–1320

[Mar+10] Mark D et al (2010) Microfluidic Lab-on-a-Chip platforms: requirements, characteristics and applications. J Chem Soc Rev 39(3):1153–1182

[Market13] Microfluidic applications in the pharmaceutical, life sciences, in-vitro diagnostic and medical device markets report 2013 (2013). http://www.researchandmarkets.com/research/z6nhg7/microfluidic

[OVB12] Oprins H, Vandevelde B, Baelmans M (2012) Modeling and control of electrowetting induced droplet motion. Micromachines 3(1):150–167

[PC06] Pan M, Chu C (2006) FastRoute: a step to integrate global routing into placement. In: International conference on computer aided design, pp 464–471

[PQ95] Parr TJ, Quong RW (1995) ANTLR: a predicated-LL (k) parser generator. Softw Pract Exper 25(7): 789–810

[PSF02] Pollack MG, Shenderov AD, Fair RB (2002) Electrowetting-based actuation of droplets for integrated microfluidics. Lab Chip 2(2):96–101

[PG07] Prakash M, Gershenfeld N (2007) Microfluidic bubble logic. Science 315(5813):832–835

[RW90] Ratner D, Warmuth M (1990) The $(n^2 - 1)$-puzzle and related relocation problems. J Symbol Comput 10(2):111–137

[Ric+06] Ricketts AJ et al (2006) Priority scheduling in digital microfluidics-based biochips. In: Design, automation and test in Europe, pp 329–334

[Rie+14] Riener H et al (2014) A logic for cardinality constraints. In: Methoden und Beschreibungssprachen zur Modellierung und Verifikation von Schaltungen und Systemen

[Rie+16] Riener H et al (2016) metaSMT: focus on your application and not on solver integration. Int J Softw Tools Technol Transfer, 1–17. metaSMT is available at https://github.com/agra-uni-bremen/metaSMT

[RBC10] Roy S, Bhattacharya BB, Chakrabarty K (2010) Optimization of dilution and mixing of biochemical samples using digital microfluidic biochips. IEEE Trans Comput Aided Des Integr Circuits Syst 29(11):1696–1708

[Sch+17] Schneider L et al (2017) Effects of cell shapes on the routability of digital microfluidic biochips. In: Design, automation and test in europe, pp 1627–1630

[Sin+00] Sinha V et al (2000) YAML: a tool for hardware design visualization and capture. In: International symposium on system synthesis, pp 9–14

[Sin05] Sinz C (2005) Towards an optimal CNF encoding of boolean cardinality constraints. In: International conference on principles and practice of constraint programming, pp 827–831

[Sin07] Sinz C (2007) Visualizing SAT instances and runs of the DPLL algorithm. J Autom Reason 39(2):219–243

[Sis+08] Sista R et al (2008) Development of a digital microfluidic platform for point of care testing. Lab Chip 8(12):2091–2104

[Sri+03] Srinivasan V et al (2003) Clinical diagnostics on human whole blood, plasma, serum, urine, saliva, sweat, and tears on a digital microfluidic platform. In: Proceedings of µTAS, pp 1287–1290

[SPF04] Srinivasan V, Pamula VK, Fair RB (2004) An integrated digital microfluidic lab-on-a-chip for clinical diagnostics on human physiological fluids. Lab Chip 4(4):310–315

[Sri+04] Srinivasan V et al (2004) Protein stamping for MALDI mass spectrometry using an electrowetting-based microfluidic platform. Optics East 5591:26–32

[Sto+17] Stoppe J et al (2017) BioViz: an interactive visualization engine for digital microfluidic biochips. In: Symposium on VLSI. ISVLSI

[SC05] Su F, Chakrabarty K (2005) Unified high-level synthesis and module placement for defect-tolerant microfluidic biochips. In: Design automation conference, pp 825–830

[SC06a] Su F, Chakrabarty K (2006) Module placement for fault-tolerant microfluidics-based biochips. Trans Des Autom Electron Syst 11(3):682–710

[SC06b] Su F, Chakrabarty K (2006) Yield eenhancement of reconfigurable microfluidics-based biochips using interstitial redundancy. ACM J Emerg Technol Comput Syst 2(2):104–128

[SC08] Su F, Chakrabarty K (2008) High-level synthesis of digital microfluidic biochips. ACM J Emerg Technol Comput Syst 3(4):1:1–1:32

[SHC06] Su F, Hwang W, Chakrabarty K (2006) Droplet routing in the synthesis of digital microfluidic biochips. In: Design, automation and test in Europe, vol 1, pp 1–6

[SHL16] Su Y-S, Ho T-Y, Lee D-T (2016) A routability-driven flow routing algorithm for programmable microfluidic devices. In: Asia and South Pacific design automation conference, pp 605–610

[Tri69] Trinder P (1969) Determination of glucose in blood using glucose oxidase with an alternative oxygen acceptor. Ann Clin Biochem 6(1):24–27

[WTF11] Wang G, Teng D, Fan S-K (2011) Digital microfluidic operations on micro-electrode dot array architecture. IET Nanobiotechnol 5(4):152–160

[Wan+14] Wang G et al (2014) Field-programmable lab-on-a-chip based on microelectrode dot array architecture. IET Nanobiotechnol 8(3):163–171

[WLH16a] Wang S-J, Li KS-M, Ho T-Y (2016) Congestion-and timing-driven droplet routing for pin-constrained paper-based microfluidic biochips. In: Asia and South Pacific design automation conference, pp 593–598

[WLH16b] Wang S-J, Li KS-M, Ho T-Y (2016) Test and diagnosis of paper-based microfluidic biochips. In: VLSI test symposium, pp 1–6

[Wan+17] Wang Q et al (2017) Physical co-design of flow and control layers for flow-based microfluidic biochips. IEEE Trans Comput Aided Des Integr Circuits Syst 37(6):1157–1170

[WLR11] Wettel R, Lanza M, Robbes R (2011) Software systems as cities: a controlled experiment. In: International conference on software engineering, pp 551–560

[Whi06] Whitesides GM (2006) The origins and the future of microfluidics. Nature 442(7101):368–373

[Wil+14] Wille R et al (2014) Revvis: visualization of structures and properties in reversible circuits. Lect Notes Comput Sci 8507:111–124

[Wil+15] Wille R et al (2015) Scalable one-pass synthesis for digital microfluidic biochips. Des Test 32(6):41–50

[WN99] Wolsey L, Nemhauser G (1999) Integer and combinatorial optimization. Wiley-interscience series in discrete mathematics and optimization. Wiley, New York

[XC06] Xu T, Chakrabarty K (2006) Droplet-trace-based array partitioning and a pin assignment algorithm for the automated design of digital microfluidic biochips. In: International conference on hardware/software codesign and system synthesis, pp 112–117

[XC07] Xu T, Chakrabarty K (2007) Integrated droplet routing in the synthesis of microfluidic biochips. In: Design automation conference, pp 948–953

[XC08] Xu T, Chakrabarty K (2008) Broadcast electrode-addressing for pin-constrained multifunctional digital microfluidic biochips. In: Design automation conference, pp 173–178

[XC09] Xu T, Chakrabarty K (2009) Fault modeling and functional test methods for digital microfluidic biochips. IEEE Trans Biomed Circuits Syst 3(4):241–253

[Xu+07] Xu T et al (2007) Automated design of pin-constrained digital microfluidic biochips under droplet-interference constraints. ACM J Emerg Technol Comput Syst 3(3):14

[YH14] Yu ST, Ho T-Y (2014) Chip-level design for digital microfluidic biochips. Int J Autom Smart Technol 4(4):202–207

[YYC07] Yuh P-H, Yang C-L, Chang Y-W (2007) Placement of defect-tolerant digital microfluidic biochips using the t-tree formulation. ACM J Emerg Technol Comput Syst 3(3):13

[Yuh+08] Yuh P-H et al (2008) A progressive-ILP based routing algorithm for cross-referencing biochips. In: Design automation conference, pp 284–289

[YYC08] Yuh P-H, Yang C-L, Chang Y-W (2008) BioRoute: a network-flow-based routing algorithm for the synthesis of digital microfluidic biochips. IEEE Trans Comput Aided Des Integr Circuits Syst 27(11):1928–1941

[ZC12] Zhao Y, Chakrabarty K (2012) Simultaneous optimization of droplet routing and control-pin mapping to electrodes in digital microfluidic biochips. IEEE Trans Comput Aided Des Integr Circuits Syst 31(2):242–254

Index

© Springer International Publishing AG, part of Springer Nature 2019
O. Keszocze et al., *Exact Design of Digital Microfluidic Biochips*,
https://doi.org/10.1007/978-3-319-90936-3

Printed in the United States
By Bookmasters